ENERGY
AND
EMPLOYMENT

Willis J. Nordlund
R. Thayne Robson

ENERGY
AND
EMPLOYMENT

PRAEGER

PRAEGER SPECIAL STUDIES • PRAEGER SCIENTIFIC

Library of Congress Cataloging in Publication Data

Nordlund, Willis J
    Energy and employment.

    Bibliography: p.
    Includes index.
    1. Energy policy--United States. 2. United
States--Full employment policies. I. Robson,
R. Thayne, joint author. II. Title.
HD9502.U52N67   333.7   79-22133
ISBN 0-03-055291-5

The views expressed in this work are those of the authors
and do not necessarily reflect the policies or views of the
U.S. Department of Labor or the University of Utah.

Published in 1980 by Praeger Publishers
CBS Educational and Professional Publishing
A Division of CBS, Inc.
521 Fifth Avenue, New York, New York 10017 U.S.A.

Printed in the United States of America

PREFACE

The authors of this study are surrounded by patient, conscientious, and extremely dedicated associates. In the scheme of things, a study of this nature is relatively small. However, we felt that the subject warranted examination, and fortunately, we convinced many others of its importance.

Our purpose in writing this book is really quite modest: we are convinced that the current debate on energy policy has to discuss, in a more structured way, the importance of the energy/employment relationships. We do not suggest that the discussion of issues is exhaustive; those seeking a rigorous economic analysis will be disappointed, but we hope that the vast majority of our readers will gain some sense of the urgency we feel in examining these issues more completely.

We would be presumptuous if we suggested that the ideas contained in this study were solely our own. Clearly they are not. In fact, few of them are new or novel. Rather, they are part of the literature and discussions that are present in some parts of the government and the private sector.

We cannot express our appreciation to every individual who has sparked an insight or idea. However, the Annotated Bibliography suggests the origin of many of our thoughts. In any case, we must note several sources of assistance in the preparation of this study.

We are indebted in the first instance to the National Council on Employment Policy for their support and encouragement throughout the preparation of this study. In particular, we heartily thank Professors Bernie Anderson, Curt Aller, and Garth Mangum for their reviews and comments on the manuscript. Greg Wurtzberg, executive secretary of the council, also provided helpful assistance on the project.

In the preparation process, numerous drafts were prepared, rewritten, edited, and typed. We want to make special note of the expert work of Gail Calder of the University of Utah's Human Resources Institute for her patience and understanding during the editing process. Final preparation of the manuscript became the responsibility of several people, all of whom performed in a professional and expert manner. We are particularly indebted to Joycelyn Daniels, Carol Morgan, and Julie Snapp for their efforts.

Last, but clearly not least, Lou Jean and Kathy have absorbed some of our frustrations and incurred some of their own. They are almost as anxious to see this project completed as were the authors.

CONTENTS

LIST OF TABLES

LIST OF FIGURES

ENERGY
AND
EMPLOYMENT

1
AN INTRODUCTION
TO THE PROBLEM OF
ENERGY AND EMPLOYMENT

Finding solutions to energy problems ranks high among the priorities of this nation. Grave concern over energy will continue for the next decade, if not the rest of the century. Much has been said about the economic consequences of an energy shortage, since energy, in its various forms, is an almost universal input to the production and consumption of goods and services in modern society. However, in our search for solutions to energy problems, inadequate attention is being paid to the employment impacts. The neglect of the employment dimension of energy solutions seems strange in a country whose record on unemployment has resulted in a substantial expansion of manpower services, with the highest funding level in U.S. history occurring in 1977.

This study was undertaken because we felt that too little is known about the particular employment effects of the various solutions proposed for the energy problem. Furthermore, we held a conviction—which further study has not dampened—that far too little attention is being paid to the entire range of questions relating to the impact that changes in the sources, uses, and prices of energy will have upon the level, location, and structure of employment in the U.S. economy. Proposed energy policies should be carefully evaluated in terms of their impacts upon the ability of the U.S. economy to achieve full employment throughout the country. Solutions may have very profound and, in some cases, devastating effects upon regions, industries, and jobs within the U.S. economy. But, if the consequences of energy solutions are correctly foreseen, their effects could benefit, rather than hurt, the economy.

To provide perspective to this study, it may be useful to identify explicitly several common threads that motivated our inquiry and structured our examination of energy/employment relationships.

First, we asked the question, Will finding solutions to energy/employment problems help or hinder achievement of the broader goal of full employment? Second, we asked, What mix of employment and training policies and programs will facilitate solutions to energy/employment problems? Third, while we generally believe that too little has been done in terms of examining and explaining energy/employment problems, it is clear that too many analysts are using inadequate data and analytical tools and are creating more confusion than enlightenment. It would be presumptuous to suggest that we can weave these threads into a consistent, totally coherent examination of the problem. We, too, must raise and leave unanswered a large number of questions. Unfortunately, data on many issues are unavailable. Furthermore, our purpose is not to solve problems, but to call attention to them in the hope of generating more research and thought.

The basic relationships between energy and employment can be grouped into two broad categories: energy-production industries and their employment components, and energy as a factor in the production/consumption processes of all industries. The first and most immediate concern is with energy-producing industries such as the mining, processing, refining, generating, transporting, distributing, and selling of energy as electric power, as liquid fuels, or as heating materials. The United States has a coal industry, a gas industry, an oil industry, and an electric power industry but no clearly definable energy industry. The discrete energy-producing industries have been defined for purposes of collecting current data on employment or sales activity, although the boundaries of these definitions are considered somewhat arbitrary. However, in analyzing the employment impact of changes in energy production, related transportation, distribution, and supply industries must be taken into account. We suggest that the energy industry has a broader employment scope than has been customarily defined (see Table 2).

Another problem we confront in trying to delineate the energy industry is that some industries simply do not exist in the data. For example, there is no standard industrial classification (SIC) code for the solar energy industry, the geothermal industry, the bioconversion industry, and several others. Many of these new industries have not yet emerged from the experimental stage, but their potential impact on energy production could be significant. The point is an obvious one. Given currently available data, it is difficult and perhaps impossible to circumscribe the energy industry precisely. If we define it to include the production and distribution of coal, gas, oil, and electric power, then total employment is approximately 4 percent of the U.S. labor force, or about 4.3 million workers. There is a degree of arbitrariness in that figure, but in a comparative sense, the energy-production sector—from an employment perspective—is a relatively small part of the U.S. economy. The second broad aspect of energy/employment relationships is the increasingly important role energy holds as a factor in production and consumption. Dramatic changes in the prices and availability of energy could, at the broadest macro level, curtail total growth in GNP or significantly alter the structure and location of industries across the United States.

At the micro level, energy prices determine how firms substitute labor or capital for expensive or unavailable energy. Everyone recognizes the importance of energy in the production process, but because of its historically low price, it has traditionally been included in a miscellaneous grouping of variable-factor costs. Labor typically constitutes the principal variable factor of production, although energy is rapidly gaining on labor's relative position. In fact, the prices of these two factor inputs determine their relative usage, and there may be some degree of substitutability between the two.

Again at the micro level, higher energy prices would influence consumption patterns and create changes in some aspects of household behavior. For example, assume that power companies adopt significant price differentials that decrease peak-load demand and increase off-peak usage. Could this affect the work schedules of secondary earners in the household?

We do not yet know to what degree the behavior of the national economy depends upon the structure and level of energy prices and upon the availability of alternative energy sources. Much more analysis of the energy problem is required before energy/employment relationships are placed into reasonably clear focus. The presumed relationship between GNP, employment, and energy suggests that problems in one area will spill over into others. What is not clear is which are the dependent and the independent variables in these causal relationships. The correlation coefficient for changes in GNP and changes in energy use over the last 20 years is about 0.98. This does not say or even suggest, however, that large changes in energy use permitted or restricted large changes in GNP. We simply cannot say that energy growth occurred because GNP grew, or vice versa. GNP, employment, and energy have all grown over the last several decades, as have population, government expenditures, the national debt, imports of foreign automobiles, and virtually every major social and economic indicator. High correlation exists between the rates of change of most of these variables, but we must avoid unwarranted causal connections. Examining the relationship between GNP and energy use in other countries suggests that there is nothing inevitable about this relationship in the United States. Several countries have per capita incomes equivalent to those enjoyed in the United

States, with vastly lower rates of energy use and lower unemployment.*

The most simple, but often overlooked, issue is that there is no shortage of overall energy sources in the long run; the term short-age relates to traditional forms of low-cost energy. The dramatic energy adjustments that will most likely occur within the next quarter century will be made with the expectation that new sources such as fusion, solar, shale, geothermal, wind, and wave power could, with anticipated technological advances, again provide the human family with abundant and relatively inexpensive sources of energy. Even the picture for fossil fuels may not be as bleak as some would have us believe. While politicians and environmentalists seem confident in calculating the exact amount of fossil fuel reserves, and offer timetables for their eventual depletion, geologists seem less certain about how much remains to be recovered of the world's oil, gas, and coal. However, the world's increasing dependence on fossil fuels (particularly, liquid fossil fuels) is very substantial, and most experts take as given the eventual depletion of these resources if current production and consumption patterns continue. There is a serious and long-run energy problem, but no one knows how severe it is, what its real dimensions are, what will happen if consumption patterns are not materially changed, when it will happen, or what the options are for solving the problem.

Why do we not know much about the problem, and why has the public response been one of skepticism and a seeming lack of concern? Why has President Carter's proposed energy policy been greeted with a less-than-enthusiastic response? Part of the answer is obvious: the energy problem is very complex and involves concepts, magnitudes, and technological implications that are neither well understood nor easily articulated by the layman. But a more important reason is that the very pervasiveness of energy in our society makes it extremely difficult to develop a coherent national energy policy. Even with spiraling prices and crisis predictions, energy-production/consumption relationships are marked by inertia or stability. Americans travel in gas-guzzling vehicles; the technology and resultant capital equipment used in electricity generation, or in high-energy industries such as primary metal smelting, prevent rapid substitution of one energy form for another or a reduction in energy usage.

Furthermore, Alfred Kahn argues that the prospects for a national energy policy are complicated because "we are a pluralistic

---

*Americans use more than twice as much energy per capita as do the West Germans, the Swedes, and the Swiss. The per capita income in these countries is comparable to that in the United States.

society, an individualistic society, and there is something faintly
ridiculous, if not totalitarian, in insisting that any collective compos-
ite of our individual aspirations and efforts should somehow fit into
a single coherent pattern."[1]

Evidence of our continuing inability to develop a national energy
policy is substantiated by a review of current laws. While this nation
has energy policies that cover virtually every aspect of energy production, transportation, and consumption, they have evolved haphazardly
over many decades and, as a consequence, are frequently inconsistent, outdated, inefficient, or inequitable. The hodgepodge of tax
laws related to the production, transportation, and consumption of
oil and natural gas simultaneously attempts to encourage production
through depletion allowances and certain tax writeoffs, while discouraging production through control of wellhead prices or sales in inter-
state commerce.

If policies were few and simple, then understanding the employment impacts would be easier. As it is, we are unlikely to develop
a totally satisfactory energy/employment strategy. Industry, in particular, requires a clear understanding of the directions embedded
in public policy. Virtually all energy facilities are extremely expensive, and no company will commit millions of dollars to a particular
type of development when a change in public policy could wipe out the
viability of the project. The relatively sudden shift in public attitudes
related to nuclear power plants illustrates the volatility of the decision-making process. Only one new nuclear power plant was ordered
in 1976, and plans for many plants were postponed or canceled. Questions of safety, environmental impact, fuel availability, and the projected need for electricity have convinced many people that the nation
does not need more nuclear plants.

Two more policy areas, coal and conservation, typify the difficulties encountered in trying to formulate a national energy plan. Almost everyone agrees that coal must play a vital role in energy production over the next two decades. However, that is about as far as
the agreement goes. Where the coal should be mined, how it should
be mined, what types of pollution control are required, what types of
taxes are equitable, and similar questions mark the multifaceted bat-
tle lines. Some states want to purchase electricity from coal-burning
power facilities in other states. But coal-burning states frequently
prefer to mine the coal and ship it out of state for power generation.
Coal must be burned somewhere, and typically, that somewhere has
fairly well-defined characteristics—it is near a reliable source of
water, has ready access to a dependable coal source, is reasonably
close to a major population or industrial center, and has acceptance
by the indigenous population (this factor is frequently assigned a low
level of importance). Policy decisions on coal production have wide-

spread employment impacts and should not be undertaken without regard to the employment situation in a particular area.

Conservation policies could also have a significant impact on employment, but, given our present state of knowledge on conservation/employment relationships, those impacts are impossible to predict.* There is a common assumption in most policy discussions that the rate of growth in energy consumption can be cut by one-third. This study, however, does not deal adequately with the conservation/employment issue partly because data are not available, and partly because conservation includes a vast and complex set of activities, the effects of which could vary immensely from one case to another. For example, a conservation program designed to reduce electric power or natural gas-heating requirements by insulating homes has a number of effects. It creates a tremendous demand for insulating materials, which, in the near term, increases the demand for energy, increases the number of insulation contractors, increases the cost of insulation and new construction, and increases the man-hours in construction of new homes and retrofitting of existing dwellings. But once the job has been completed, demand for insulation will presumably decline dramatically, with effects on energy conservation. In this case the short-run effects, for, say, five years, are very different from the long-run effects.

Building fuel-efficient cars means building smaller, lighter cars that presumably use less material and fewer man-hours. However, countereffects may occur if reducing car size and weight results in an increased demand for trucks, vans, and buses that can move larger numbers of people. Smaller cars generally require more hours of maintenance and service and thus may shift employment from manufacturing plants to repair shops. The net effect on the use of energy is not entirely clear. For this reason, the employment impacts of conservation policies and practices are largely ignored in this study, even though these issues may be among the most important in the linkages between solutions to energy problems and employment.

While there is a severe deficiency in the analytical capabilities not only of the conservation program, but of the entire area of emerging alternative technologies, some work has been done. We have attempted to delineate some of the current issues and summarize several of the preliminary research efforts. The scarcity of material in this area of activity suggests a potentially fruitful avenue of intensive inquiry.

---

*To date, the Center for Advanced Computation, at the University of Illinois, has provided most of the methodological and empirical work in this area. A summary of some of this work appears in Chapter 6.

This study focuses attention on the types of issues U.S. policy makers, industrial leaders, and researchers should consider in a rational examination of energy/employment relationships. There is no clear dichotomy by energy type, economic category, time period, or geographic location, because the pervasiveness of energy usage prevents a clean compartmentalization of the analytical structure.

Following the basic division discussed earlier with regard to energy/employment relationships, the next three chapters focus on energy-production industries and their employment components, while Chapter 5 deals with energy as a factor in production and consumption.

Chapter 6 is a summary of the employment dimensions of alternative technologies.

Useful projections cannot be made without better energy/employment models. Therefore, three models are reviewed in Chapter 7. In addition, to give some continuity to the exploration of energy/employment problems, we develop a research agenda and make some energy/employment policy recommendations in the last two chapters.

NOTE

1. Alfred Kahn, "Impact of Regulatory Procedures on the United States Energy Condition," in A Sensible Energy Policy Now: Today's Challenge to Meet 21st Century Needs, National Energy Forum 5 (Washington, D.C.: U.S. National Committee of the World Energy Conference, May 23-24, 1977), pp. 8-9.

2
EMPLOYMENT IN
THE ENERGY SECTOR

It is useful to think of the energy-producing sector as two distinct industrial groups: the established industries and the emerging industries. In the former group are coal mining, oil and gas extraction, electric power generation, and light-water nuclear reactors; in the latter group are solar, geothermal, oil shale, bioconversion, breeder reactors, thermal gradients, and nuclear fusion. This distinction is important because, in terms of the employment impact over the next two decades, virtually all expansion will occur in the established energy industries. Every emerging energy industry is extremely capital intensive, and until massive resources are committed to those activities, their employment impacts will be minimal. Therefore, this chapter deals with employment in the established energy industries.

As noted earlier, there is no precise definition of the energy sector. Consequently, it is very difficult to quantify the character of the energy work force. The following discussion estimates the dimensions of the energy-sector work force, but we readily admit that most of these figures can be challenged. Our purpose, however, is to outline the general manpower characteristics of the energy segment of the economy, to suggest what determines the supply and demand patterns within this segment, and to describe several employment-estimating techniques.

EMPLOYMENT IN THE ENERGY SECTOR, 1973

The 1974 Labor Report of Project Independence suggested that "labor employed in the design, construction, operation, and maintenance of energy production facilities accounted for slightly more than 2 percent of the nation's total work force in 1973."[1]

However, it must be recognized that there are dimensions of the energy sector not accounted for in this estimate. For example, environmental monitoring and control, the entire regulatory process, and research and development activities are conspicuously absent. In addition, all of the secondary employment impacts in industries supporting the energy sector, such as steel and wire production and fabrication, transportation, and myriad support services, would expand these estimates by two- or threefold.

In any case, the employment generated by the total energy sector is sizable and is certain to expand rapidly. As a consequence, human resource planners at all levels are confronted with a major challenge. Some sources of labor supply will be impacted to a much greater extent than others. The technologies inherent in the energy sector require a disproportionate number of highly skilled professional and technical employees. The relative importance of foremen and skilled craftsmen is even more noticeable.

As Table 1 shows, about one-third of all energy-production employees are in craftsmen and foremen categories, while 13 percent

TABLE 1

Distribution of Production Employment by Major Occupational
Classifications
(in percent)

Occupational Classification	Energy-Production Industries	All Industries
Professional, technical workers	13	14
Engineers	4	1
Managers, proprietors	6	10
Sales workers	1	6
Clerical workers	19	17
Craftsmen, foremen	32	13
Operatives	21	17
Service workers	2	13
Laborers and farm workers	5	9

Source: Federal Energy Administration, Labor Report, Project Independence Blueprint, Final Task Force Report (Washington, D.C.: U.S. Government Printing Office, November 1974), table II-3, p. 19.

TABLE 2

Employment in the U.S. Energy Sector, 1976
(in thousands)

Industry	Employment
Coal mining	214.3
Construction	
Nuclear	60.1
Hydro	8.4
Fossil	57.9
Electrical work	325.4
Transportation	
Railroad (20 percent coal revenues)	105.5
Truck (12 percent coal revenues)	121.0
Barge (15 percent coal ton-miles)	29.6
Pipelines	16.7
Gas and oil extraction	360.3
Electric companies and systems	314.9
Gas companies and systems	159.5
Electrical industrial apparatus	212.3
Petroleum refining	157.1
Heating equipment exelectric	37.9
Steam engines and turbines	42.9
Construction and mining equipment	337.0
Power-transmission equipment	47.2
Transformers	48.0
Electrical lighting and wiring equipment	195.4
Electronic components and accessories	372.0
Railroad equipment (20 percent coal revenues)	8.6
Other petroleum and coal products	45.8
Electrical goods	319.1
Gasoline service stations	627.0
Total	4,223.9

Sources: Data for all industries except construction are from U.S., Department of Labor, Employment and Earnings 24, no. 3 (March 1977): 61-69. Construction estimates are derived from the Construction Manpower Demand System, U.S. Department of Labor.

of all workers fall in these classifications. Similarly, engineers account for about 1 percent of all workers, but 4 percent of the energy sector's work force. Only 3 percent of the energy work force is in sales and service occupations, compared with 19 percent of the total work force. The Federal Energy Administration (FEA) noted that "this high proportion of workers in unskilled occupations reflects the technical and highly capital intensive nature of energy production."[2]

These relationships suggest what sources of labor supply are likely to be most important as energy production expands. First, because of the high proportion of craftsmen and foremen needed, apprenticeship programs are going to be extremely important. Unions and companies through their joint apprenticeship councils must be responsive to these needs by anticipating the number of craftsmen required four or five years in the future. Second, secondary and postsecondary schools must be aware of the expanding occupational opportunities for clerical workers and graduates trained in petroleum, mining, chemical, and mechanical engineering or in other engineering specialties.

The FEA estimate that placed about 2 percent of the nation's work force in the energy sector was conservative. As we noted, the problem of defining the energy sector hinges on deciding which industries to include. Further problems arise in determining what segment of the transportation and capital-producing industries are part of the energy industry. Despite this difficulty, we have attempted to isolate the industrial categories generally thought to be directly associated with some phase of energy production, transportation, and consumption. Several of the categories listed in Table 2 may be marginal, but the estimate is still relatively conservative. Based on the list in Table 2, employment in the energy sector is nearer 4 percent of the work force than 2 percent.

THE LABOR INTENSIVENESS OF VARIOUS ENERGY SOURCES

Developing an energy/employment strategy can take several alternative points of departure. For example, we could ask how energy shortages affect production with consequent impacts on employment. Or, we could calculate how many workers the various energy industries require to supply national energy needs. An important extension of the latter approach is to ask a somewhat different question: How much employment is generated by different methods of energy production? In other words, which source of energy production is most labor intensive?

TABLE 3

Quads of Energy by Source, United States, 1976

Source	Quads of Btu's[a]	Percentage of Consumption
Oil	34.9	43.9
Domestic supply	19.9	
Supplemental supply[b]	15.0	
Natural gas	20.2	30.6
Coal	13.7	18.6
Electricity (nuclear-hydro)	5.1	6.9

[a]Quadrillions (10^{15}) of British thermal units.
[b]Supplemental supply includes imports, shale oil, and coal liquefaction.

Source: Bureau of the Census, Statistical Abstract of the United States, 1977 (Washington, D.C.: U.S. Government Printing Office, 1977), p. 594.

Tables 3 to 5 examine energy production from various sources and translate energy output into ratios of direct employment to quadrillions of British thermal units (Btu's), in each of the production sectors. This approach gives an estimate of the relative labor intensiveness, on a Btu basis, of various forms of energy production.

Table 3 shows the basic sources of U.S. energy by major category. These estimates show the relative importance of each energy source in terms of its contribution to the economy. This is not to suggest that the various energy sources are equivalent. A quad of energy from natural gas may be preferred to a quad of energy from coal or nuclear power. These relative characteristics depend on current technology, relative availability, environmental standards, and so forth.

Employment in each of these industrial groups in 1977 is shown in Table 4.

In terms of direct employment undifferentiated by occupation, location, or skill level, as Table 5 shows, the electricity (nuclear-hydro) source is the most labor intensive. This runs contrary to most thinking and therefore requires brief comment.

If we examine the subindustry classifications in Table 4, it is obvious that the electricity-distribution network accounts for more than three-fifths of total employment. No comparable category can

TABLE 4

Employment in Energy-Production Industries in 1977

Energy Source	Employment
Oil	
Oil and gas extraction*	202,250
Oilfield machinery	72,200
Petroleum refining	160,300
Other petroleum and coal products*	24,500
Total	459,300
Gas	
Gas companies and systems	157,600
Oil and gas extraction*	202,250
Total	359,800
Coal mining	217,500
Other petroleum and coal products*	24,550
Total	242,050
Electricity	
Steam engines and turbines	42,600
Power-transmission equipment	47,800
Transformers	50,200
Electric companies and systems	314,900
Total	455,500
Construction	
Nuclear	63,600
Hydro	7,300
Fossil	59,100
Total	130,000

*These figures are divided by two, since separate data are not given in Employment and Earnings.

Sources: Data for all industries except construction are from U.S., Department of Labor, Employment and Earnings 25, no. 3 (March 1978): 62-69. Construction estimates are derived from the Construction Labor Demand System, U.S. Department of Labor.

TABLE 5

Employment by Major Energy Source, 1976

Source	Quads of Btu's	Employment	Employment/ Quads
Oil	34.9	459,300	13,160.5
Gas	20.2	359,800	17,811.9
Coal	13.7	242,050	17,667.9
Electricity	5.1	455,500	89,313.7

Source: Compiled by the authors from data in Tables 3 and 4.

be easily defined for oil and coal. Comparisons of this type invariably run into problems of how the industry is defined, what secondary and tertiary elements should be counted, and how multiple products should be counted. Therefore, for purposes of analysis, the nuclear-hydro classification is probably not comparable with the other three.

The oil, gas, and coal industries have similar employment/quads ratios. In fact, it would be very difficult to suggest an employment strategy based on their differences. To focus on this relationship, it is necessary to define more precisely each industrial category, the employment associated with it, and its relative contribution to the economy's energy output.

It seems reasonable to suggest that the basic technology and scale of operation of energy-production industries are typical of capital-intensive processes. No energy industry is very labor intensive. Therefore, energy industries may have more of an effect on occupational and geographical distribution of employment than on national employment/unemployment rates.

ENERGY LABOR DEMAND

What determines the demand for labor in the energy sector? In the energy sector, as in all industries, the demand for labor is a derived demand. In other words, labor is a factor input to the production process, similar to capital, raw materials, and land. There must be a demand generated for a product (in this case, various energy forms), which in turn dictates what types and quantities of factor inputs are required. Embedded in the decision-making process are a variety of exogenous variables, such as technological, environmental, economic, and political considerations. For example, the product demand may be specified as 650 million tons of coal per year. There are several ways to obtain the required coal output, each of which has different employment implications. If, due to environmental requirements, low-sulfur coal is preferred, western lignite will be the probable source. Lignite production requires strip mining, which has an occupational structure using heavy-equipment operators, electricians, blasters, truck drivers, mechanics, and a variety of other workers. If, on the other hand, demand is strong for high-Btu bituminous coal, underground sources will be preferred. Underground mining requires a very different array of workers, such as continuous mining machine operators, electricians, roof bolters, shuttle-car operators, and mine inspectors.

Similarly, if the demand is for 100 million kilowatt hours of electricity, several technologically different sources can be used. Coal-fired steam plants, nuclear plants, hydroelectric facilities, geothermal plants, and a variety of more esoteric electricity sources are typical examples. Political and environmental pressures may discourage nuclear power plant construction, in favor of generating more electricity from coal-fired plants. The type of operating and maintenance employees in a coal-fired plant differ in training and number from those in a nuclear facility. Furthermore, a coal initiative expands employment in the coal mining industry, whereas emphasis on nuclear power enlarges employment in uranium mining and processing.

Another example of how energy alternatives impact upon employment is found in the oil industry. A demand for 5 million barrels of oil per day can be fulfilled in a variety of ways, each with different employment components. If Americans opt for production of natural crude oil, employment in occupations related to oil exploration and well drilling will grow. If coal conversion is preferred, employment in coal mining and coal-conversion plant operation and maintenance will increase. If oil imports are the desired option, employment will grow in occupations related to shipping or to the construction and maintenance of pipelines and oil-storage facilities.

However, the process of estimating employment demand does not stop when the option has been selected and production started. The energy industry is not a self-contained production entity. The examples listed above only suggest what the direct employment requirements will look like. To assess the total employment impact, the analyst must determine how a selected energy option affects secondary or indirect employment changes as well. For example, if coal conversion is selected, the employment of welders, pipefitters, electricians, millwrights, machinists, and a variety of other occupational groups will increase immediately. In addition, the development of the coal-conversion industry will expand output in the steel-pipe industry, the high-pressure valve industry, and myriad raw-material and fabricated-parts industries. Each output change would affect employment in these secondary industries.*

To finish off the analytical framework, we must attempt to estimate the nature of employment changes that will occur in support of the basic industrial changes. As the basic industry develops in a particular area, more workers and their families will move there. A population change generates increased demand for the full spectrum of goods and services, with specific employment implications. This third category of change is frequently referred to as the tertiary impact.

It may not be immediately obvious why tertiary employment changes are of concern to the analyst, but there are several principal reasons. Construction workers can build a coal-conversion plant or they can build homes, schools, or hospitals. To the extent that the coal-conversion project attracts local construction workers, a deficiency may develop in the local supply of construction workers for nonenergy projects; the following section, on socioeconomic impacts on the labor market, outlines the implications of these processes.

In addition, the tertiary sector is frequently characterized by a predominance of service and light-manufacturing industries. Wages and salaries in these industries are typically lower than in the construction trades. As a consequence, either workers will be drawn away from these industries in response to the income differentials, or there will be major wage increases to stabilize employment. If

---

*Primary, secondary, and tertiary employment do not represent precise analytical tools, but they are useful in providing insights into the mechanics of the labor market. They pertain to the industrial perspective of employment as opposed to the individual-worker perspective. They are not, for the most part, macroeconomic variables. Primary employment typifies the number of workers engaged directly in a well-defined production process. Because of the interrelatedness of the U.S. economy, virtually all production activity requires inputs from a variety of support or secondary industries. The employment generated in these secondary industries is secondary employment. It is, in a sense, once or twice removed from the primary production process. Tertiary employment encompasses the myriad workers who provide goods and services to workers and their families in the primary and secondary sectors, but are not, in any meaningful sense, functionally related to the primary production process. A large part of the service sector would typically fall in the tertiary sector.

the former occurs, the quantity and quality of goods and services flowing from the tertiary industries may decrease, resulting in local inflationary pressures.

As we have pointed out, the estimation of energy labor demand does not stop at an energy facility's front gate. There is a complex series of repercussions that extends to the local economy, and in most cases, to industries and areas far removed from the development site. These interactions do not give the analyst the freedom to restrict his analysis to primary employment even if he were inclined to do so.

The estimation of direct employment requirements is a relatively simple process, requiring a knowledge of what type of energy facility will be built (underground versus surface mining), where it will be located, what size it will be (say, 5 million tons per year, 1,000 megawatts), when it will be initiated, and how fast construction will occur. If it is known that on July 1, 1978, construction on a 1 million-ton-per-year underground coal mine will be started at a town in Colorado, with coal production to begin January 1,1980, fairly precise employment estimates for the construction, operation, and maintenance of the mine can be made.[3] The Bureau of Labor Statistics, the Bureau of Mines, and a variety of state agencies maintain updated employment tables for most major energy industries. These tables reflect or implicitly include changes in technology.

A standard technique for estimating the indirect and tertiary employment changes is to apply a reasonable employment multiplier to the direct estimates. These multipliers typically range between one and three. However, there is very little justification (other than informed judgment) for selecting a specific multiplier in a specific situation. The analyst may attempt to locate a similar development situation, ascertain what happened there, and then apply a similar multiplier to the case in point. While there is some intuitive appeal to this procedure, it is a very risky undertaking. It becomes increasingly hazardous as the size of the project increases or the size of the labor area decreases.

For example, if the analyst wanted to know how many total jobs would be created if firm Z moves into Denver, and firm Z has 73 employees, a reasonable multiplier of 2.0 or 2.5 could be applied. Whether the total eventually turns out to be 146 or 183 probably does not make a lot of difference in the Denver labor market. The analyst would probably not be fired if his estimate were very high or very low, because it is doubtful that anyone could determine precisely what the impact would be anyway. The ad hoc multiplier is a relatively safe procedure in large labor market areas.

On the other hand, suppose a 50,000-barrel-per-day coal-liquefaction plant is being built in a town in North Dakota, population 129.

The construction work force would peak at 3,000 workers in the third year of construction. A permanent work force of 1,400 would eventually operate the plant. Now, the analyst must select an employment multiplier that will help community planners anticipate the impact. It definitely makes a difference whether he selects 1.5 as opposed to 2.5. In fact, it may be reasonable to suggest that if he selects 2.5, and the correct multiplier turns out to be 1.5, his reputation as an analyst will be ruined. The difference will be noticed, and the impact-abatement plans will be meaningless. What compounds the problem is that a reasonable population multiplier of, say, 2.0 to 2.5 must be applied to the total employment change. Suppose the analyst applied a multiplier of, say, 2.0. Total population was therefore projected to increase to 7,129 ($1,400 \times 2.5 = 3,500 \times 2.0 = 7,000 + 129 = 7,129$). Based on this estimate, roads, homes, schools, hospitals, and other public and private facilities were built. If in fact the employment multiplier were 1.5, employment would increase to 2,100, with total population increasing to 4,329 ($1,400 \times 1.5 = 2,100 \times 2.0 = 4,200 + 129 = 4,329$). Community planners would not be particularly happy about the 39 percent excess capacity in their schools, hospitals, and other facilities. Therefore, the reasonable employment-multiplier approach is an extremely risky procedure in small labor markets.

The analyst who uses a naive or simplistic approach is often justifiably criticized. It is important, however, to point out that the pressure exerted on analysts to come up with a set of numbers is tremendous. Planners use these numbers to justify budget proposals, convince industry about anticipated growth, or seek external funding. In many cases, planners are not concerned about where the numbers come from, as long as they can be justified in some manner. In-the-ballpark estimates, it is argued, are better than no estimates at all.

The Size of the Employment Multiplier

It is possible to suggest why some areas are likely to have large employment multipliers while others have relatively small ones. The first determinant is the degree to which the labor market area becomes self-sufficient. In other words, the more the community must provide for itself, the more workers it will require. Therefore, the degree of geographic remoteness is a critical variable. In addition, the existence of reliable transportation may permit long-distance commuting and therefore reduce the indigenous labor market requirement.

A second related factor is the degree to which trade flows evolve between the community and other economic areas. The more income that leaks out of the community, the smaller the employment multiplier. This factor is proportionate to the self-sufficiency and remoteness of the community.

Third, the degree of permanence also influences the ultimate size of the employment multiplier. As noted previously, the influx of support workers (the secondary and tertiary buildup) is likely to be relatively greater for a work force of 1,000 working in coal mining than it would be for a comparable increase of construction workers with an expected tenure of three or four years.

In addition to these factors, there are a number of peripheral determinants that may or may not be important in a particular location. Examples are zoning laws, the existence of an educational system, the availability of skills in the local economy, and public attitudes and acceptance of outsiders.

One of the most difficult problems with multiplier analysis is that it is relatively inflexible and, to a large extent, static. Most techniques permit the estimation of a beginning point and an ending point, but of nothing in between. In other words, we estimate that employment is X today and 2X 24 months hence, but are hard pressed to explain the dynamics of how or where the change will occur.

The importance of this deficiency cannot be overstressed. It makes a difference, in terms of local labor market planning, to know whether 90 percent of the expansion will occur within the first three months or the last three months of the period. In fact, to ensure a smooth transition to a larger local labor force, planners may want to change the timing of the buildup so that population increases tie in with output from the educational system, the phaseout of large projects in an adjacent community, or some other relevant labor market change. Current analytical tools do not, in general, permit analyses exhibiting this degree of precision.

Flexibility Is the Key

Estimating energy labor demand at the local level is an exceedingly sensitive process because small errors multiply, affecting many people. In most cases, however, local area estimates are a one-shot process. The situation is different at the state and national levels. Policy options evolve in response to changing economic, technological, political, environmental, and international conditions. Since labor market conditions react to alternative policy options, it is important to develop a flexible labor demand-estimation technique. The volatility of the domestic and international economic and political relationships suggests that predicting future policy initiatives is very risky. This, of course, is precisely the dilemma firms in the energy industry face without the existence of a federal energy policy. They are very reluctant to commit large capital resources to an energy venture that the federal government eventually may not support.

As noted earlier, there is no single analytical technique that will provide all the answers. However, the complexity of the energy-development process, in conjunction with the uncertainties of federal and state policy directions, strongly suggests that an ongoing analytical program be developed and maintained.

ENERGY LABOR SUPPLY

One of the most difficult problems economists continually encounter is the estimation of labor supply (see the Appendix for a more complete discussion of labor-supply issues related to energy development). Contrary to the rather limited number of variables determining the demand for labor, there are undoubtedly dozens of variables, both quantitative and qualitative, that affect the supply of labor. In addition, the demand for labor is relatively site specific. In other words, it is relatively easy to determine where the job openings are that constitute the demand function, both geographically and industrially. Firm X has Y job openings in occupation Z at wage rate W, during time period T. However, attempting to determine who will fill specific job openings is a vastly more difficult problem.

There is an implicit assumption in the academic community, the business sector, and governmental sector that if the demand for labor increases in a specific area, the supply of labor will somehow flow to meet it. However, it is unclear how long the adjustment process will take, where the workers will come from, what actions firms will take if there are insufficient numbers of a particular type of worker, what the cost (both direct and indirect) of labor mobility is, what can be done by established institutions (government, unions, educational institutions, and firms) to improve the adjustment process, and so on. It does seem reasonable to suggest that if enough time is permitted in the adjustment process, equilibrium will be approached in some sense. But, it is also reasonable to assume that if the adjustment process is fairly long, it will adversely affect the firm and community. Increased cost, reduced production, intensified recruitment efforts, and so forth will hurt the firm. The community loses real output, experiences higher rates of inflation, and utilizes its resources less efficiently. In a conceptual sense, all of these costs will occur if the adjustment process is not instantaneous. However, as a practical matter, we tolerate a less-than-efficient labor market. In fact, these costs are part of the price Americans pay for a free enterprise system where workers have the freedom of where to work or whether to work at all.

Nevertheless, to utilize the nation's resources more productively, efforts should be made to facilitate the adjustment process. Several mechanisms that attempt to improve the functioning of the labor market are used by the U.S. Employment Service, unions, and university placement offices, but market conditions in the energy industry are different enough to make established mechanisms only partially useful. Several examples will suggest the practical differences. First, energy development frequently occurs in relatively remote areas with small indigenous populations, while the established mechanisms function principally in densely populated areas. (The rationale for this situation has historical and economic underpinnings.) Second, the pace of development is destined to be very rapid. It is difficult for labor-supply sources to respond to this moving target. Third, the emergence of new industries (oil shale, coal conversion, geothermal, solar) and new legislation will result in the emergence of new occupations (particularly in the areas of environmental protection and restoration and facility maintenance). Fourth, in this development process, reliance on in-migration is going to be extensive. Traditional mechanisms simply are not capable of accurately estimating the labor-mobility phenomenon. The Employment Services's interstate-clearance process does not even scratch the surface of the problem.

The boom conditions related to energy development are not unprecedented, but their magnitude and duration may be unique in the American experience. The construction of missile bases in northern North Dakota, Cape Canaveral in Florida, the pipeline in Alaska, and the aircraft industry in Washington are contemporary examples of boom situations. Each covers either a relatively small geographic area or industrial sector, or both. In contrast, energy development in the Rocky Mountain region covers a very large geographic area, a diverse industrial sector, and a long period of time. The labor-supply requirements for this process are not clearly understood, and at present, no mechanisms exist to provide reasonable, accurate, ongoing information.

The next chapter discusses energy/employment issues as they relate to boom regions.

NOTES

1. Federal Energy Administration, Labor Report, Project Independence Blueprint, Final Task Force Report (Washington, D.C.: U.S. Government Printing Office, November 1974), p. 14.

2. Ibid., p. 19.

3. These estimates may turn out to be inaccurate, however. For example, the Jim Bridger Power Plant, in Rock Springs, Wyoming, eventually employed more than twice as many construction workers as the number projected. See Joe G. Baker, "Case Study 2: Sweetwater County, Wyoming," Labor Allocation in Western Energy Development, Monograph no. 5 (Salt Lake City: University of Utah, Human Resource Institute, 1977), p. 36.

3
SOCIOECONOMIC IMPACTS OF
ENERGY DEVELOPMENT

Many of the reports analyzing energy/employment relationships have centered on development in the Rocky Mountains—northern Great Plains region. Large-scale energy projects, with all of their potential for disrupting both the social and natural environment, are very prominent among the economic activities of this resource-rich area. Most of the studies have dealt with power plant construction, operation, and maintenance, and with coal mining (both surface and underground). One of the most important labor market observations that comes out of these reports is that at least in this region, a chaotic community causes a chaotic labor market, and not vice versa.

Planners and some local government officials have a tendency to blame a multitude of social ills on the labor market mechanism. However, case studies reveal that the principal reason for labor market imbalances is the poor quality of life the community provides. Workers relocate for high-paying jobs but will not tolerate the lack of goods and services that their incomes and prior life-styles have led them to expect. To human resource specialists this may seem like a trivial point, but to researchers in the field who are constantly told that labor market adjustments are the major barrier to the development of a stable work force, this is extremely important. In case after case, initial adjustments did occur, but adverse socioeconomic conditions caused high turnover, low productivity, and absenteeism. High wages attract workers to a site, but a suitable environment and a reasonably comfortable life-style are necessary to keep them there.

THE BOOMTOWN SYNDROME

The energy-development process is unlikely to be orderly or efficient. In fact, there is evidence that energy development disrupts a broad spectrum of economic, political, social, and psychological relationships in the host community. The boomtown syndrome is not a new phenomenon of U.S. life, and much is known about it. Breakdowns occur in the local political structure as outsiders move in and dominate the political machinery; crime, alcoholism, family dissolution, and drug use increase; schools and hospitals become overcrowded, and inadequate roads, water, and sewer systems impose hardships on the community's residents. All of these conditions are symptomatic of the rapid influx of people in a boomtown.

Without going into a great deal of detail about why boomtowns develop, it may be useful to summarize the typical process. Boomtowns typically begin as small communities, geographically separated from large urban areas. In many cases, the community is relatively stable in the sense that it has an established political structure, adequate infrastructure, closely knit social relationships, and so forth. Some small communities may be experiencing slow population decreases as young people migrate to the cities. (Residents frequently welcome large-scale energy development as a means of providing jobs for young people and keeping them in the community.) Also typical of the community are adequate, but old, physical facilities, such as water and sewer lines, waste-treatment plants, schools, and hospitals.

Then one day something unusual occurs. Someone discovers a deposit of low-sulfur bituminous coal closeby and decides to build a large power plant. A few strange faces appear in local hotels as an advance team surveys the prospective site and makes preparations to begin development. These individuals find a tightly knit community where everyone knows everyone else on a first-name basis. Few assimilation problems occur during this phase. The number of outsiders is small; they are viewed as temporary; and their demands on the system are modest.

Several months after the arrival of the surveyors, some trucks roll into town with bulldozers and other heavy equipment. In addition, a substantially larger number of new people come into town. The boomtown process is under way.

Community leaders suddenly realize that what is happening is not a temporary phenomenon—it will not go away if it is ignored; and, in fact, it will probably change the entire community's economic, political, and social character. Concerned community representatives contact the company and raise questions about population growth, wage and price changes, and tax increases. At this early stage, many of the noneconomic problems have not occurred. Precise answers are virtually never provided, because the company's ability to predict is not good, or because in some cases the company considers the information proprietary. As a result, the company is very circumspect and elusive, and the community leaders become frustrated.

Within several months a large contingent of construction workers and their families move into the area, further disrupting the community. Increasing demands are made on the water, sewer, gas, and
electricity systems. Roads deteriorate, to the frustration of both
long-term residents and newcomers. There are not enough class-
rooms in the schools to accommodate the new students; the hospital
is too small. Vandalism, vice, and public intoxication increase.

This brief scenario of the boomtown process suggests several
of the major causal factors. First, development is frequently unex-
pected, or at least the magnitude of change is not known. Second,
change is very rapid. Third, firms are frequently reluctant to divulge too much information, for proprietary and other reasons.
Fourth, there is essentially no planning capability at the community
level. Fifth, the need for social overhead capital and the full range
of social services arises before revenue flows develop to pay for them.
This has become known as the "front-end" financing problem. Sixth,
social frictions spring up between the indigenous population and outsiders. Seventh, prices and taxes rise rapidly. This problem places
particular hardships on those local residents who are relatively im-
mobile, do not directly benefit from the new industry, or are on some
form of fixed income.

Therefore, it is important to recognize that the impacts of rapid
energy development do not accrue uniformly for all those in or near
the impact area. Different segments of the community experience
different costs, most of which they cannot directly control. For example, the elderly on fixed incomes are directly damaged by rising
prices, taxes, crime, and deteriorating roads, hospitals, and parks.
These residents may have few options, such as leaving the community,
seeking gainful employment, or protecting themselves against crime.
The unskilled and poor residents probably were experiencing difficult
economic conditions before the boom, and depending on their ability
to gain employment at the energy plant site, many experience more
intense hardships or improve their condition materially. Small businesses may, at least initially, welcome the growth of business. However, as new, more efficient competitors move in, the established
firms may experience increasing difficulty. Even more important in
the long run may be the overbuilding of businesses, so that excess
capacity exists as the construction phase winds down.

Why is not more done to prevent or mitigate the problems boomtowns face? One reason is that government officials at the state and
federal levels are reluctant to intervene in local affairs without the
sanction of the local population. In fact, many communities believe
they can handle the problem and therefore reject external assistance.
Also, community officials may hear rumors of rapid increases in development, but they will be reluctant or unable to enlarge utility sys-

tems, schools, and other public services until construction actually begins and the tax base enlarges. However, the lead time on many social-infrastructure elements is several years. Therefore, the need arises much faster than the ability to cope with it. Furthermore, the absence of a local planning capability, or the mistaken belief that planning is not needed, results in ad hoc responses that are inconsistent, poorly timed, inefficient, and inadequate. The "ostrich syndrome" is very difficult to resolve, and valuable time is lost in responding to the mounting problems.

Everyone would agree that advanced information and planning are absolutely critical in preventing the onslaught of boomtown problems. However, other problems still exist, and their resolution is extremely difficult. An important dimension of the community-impact problem is that there is likely to be a major difference in the short-run (three to five years) and the long-run (20 to 25 years) requirements of the population. To illustrate the problem, consider the population impact during the construction phase of a 2,250 megawatt, coal-fired

FIGURE 1

Employment and Population Added by Construction
(2,250-megawatt, coal-fueled, electricity-generating plant)

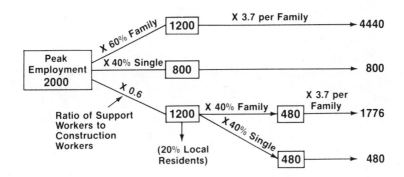

Source: U.S., Department of Housing and Urban Development, Rapid Growth from Energy Projects: Ideas for State and Local Action, HUD-CPD-140 (Washington, D.C.: Government Printing Office, April 1976), p. 5.

FIGURE 2

Employment and Population Added by Operations

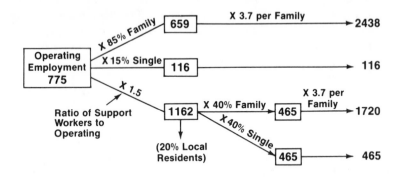

Total Population Added 4739

Source: U.S., Department of Housing and Urban Development, Rapid Growth from Energy Projects: Ideas for State and Local Action, HUD-CPD-140 (Washington, D.C.: Government Printing Office, April 1976), p. 6.

electricity-generating plant (see Figure 1). The example is hypothetical, but it illustrates the basic problem.

Peak construction employment of 2,000 workers has resulted in a population increase of almost 7,500 people. If this increase were predicted to last for 25 to 30 years, local planners could easily figure out what social overhead capital would be needed and proceed to put the facilities in place. The problem is that after the employment peak, there will be a rather rapid decrease toward zero. Eventually, all construction work will cease and the facility will enter the operation and maintenance period. A similar schematic will show the changes likely to occur in employment and population during this period (see Figure 2).

Initial observation may suggest that the difference between 7,496 and 4,739 is not that large. A large metropolitan area could easily absorb a population increase of this size, and, in fact, may welcome it. Small communities have different problems. First, how does the community accommodate the construction population during the three to five years they are in the area? They need and deserve services

as much as the indigenous residents. Second, the operation and maintenance work force and population is typically different from the construction work force and population; their mix of skills, their attitudes
toward the community, and the types of services required differ substantially. It may not be easy to make the transition from one contingent of new residents to the other.

In some sense, the resolution of these problems involves value
judgments about what different types of community residents deserve
and need. What is clear is that these are extremely important issues
that can be the key to what holds together a community under stress,
or tears it apart.

The Generalized Character of Impacts

Labor markets in boomtown areas are profoundly sensitive to the
boomtown environment. More specifically, labor market problems in
these areas cannot be adequately understood without a clear understanding of the complex relationship between work and the nonwork
environment. U.S. workers are sensitive to wage rates, and traditional economic theory suggests that wage-rate differentials will
prompt mobility. However, it is a well-known fact that workers respond to noneconomic incentives or disincentives as well, and, in
fact, in many cases the noneconomic factors are of major importance.
High wage income only partially compensates for negative noneconomic
factors, such as a dirty or unsafe work environment, excessive hours,
long commuting distances, insensitive supervision or management,
and temporary or irregular work assignments.

Most energy-development projects are, by their very nature,
massive undertakings. They involve hundreds of millions of dollars,
thousands of workers, and long periods of time for completion. In
addition, they are frequently located in geographically remote areas.
To attract a work force to these remote regions (the indigenous labor
force is never sufficient), high wages are used. In most cases, this
carrot is enough to lure an adequate labor force to the area.

However, high wages do not ensure retention of workers, high
productivity, or regular attendance. Labor turnover is related to
productivity; absenteeism is related to turnover and productivity. In
fact, all three factors form a complex response pattern to conditions
that are both internal and external to the work environment. The
character of this pattern has major policy implications related to en-
suring sufficient labor for major energy initiatives.

As large numbers of workers arrive, housing shortages, inadequate hospital space, traffic congestion, and increased crime cause
social well-being to deteriorate. As problems arise, the indigenous

population becomes increasingly irritated with the outsiders. The latter sense the negative attitude, which impedes constructive interaction between the groups. The focal points of initial disruption are the frustrated expectations of new residents for a good life and the breakdown of the established social patterns of the indigenous population.

Unfortunately, workers are unable to separate their work and nonwork life. As a consequence, attendance falls and tardiness increases. To offset this problem, firms may hire additional workers. The increased labor force further aggravates the shortage of housing and services.

Since many of the problems in the community take years to correct, some workers simply give up and quit their jobs. Many move out of the area. This turnover problem has several dimensions. First, on the presumption that the firm hired the best employees available initially, subsequent workers are likely to be less productive, that is, younger and less experienced. Even for those workers with the potential to become valued employees, a break-in period of relatively low productivity is still necessary. Reduced productivity slows the development process and raises costs. Second, turnover at the construction site typically siphons the supply of workers in the local community who are attracted to higher-paying opportunities. Accordingly, shortages of workers in non-energy sector jobs result in reduced or eliminated services and increased costs. The deterioration of services further impacts the quality-of-life problem in the area.

While major companies can anticipate the levels and timing of raw materials, labor, and capital needed for any large construction project, they have a harder time estimating variables that affect productivity. To gain a further appreciation for these types of problems, the energy boom in Rock Springs, Wyoming, is discussed below. In this case, the principal firms failed to consider a variety of variables that ultimately had a significant impact on their ability to acquire and retain workers.

A LOCAL LABOR MARKET UNDER STRESS: ROCK SPRINGS, WYOMING

Rock Springs is a relatively small, isolated community 170 miles east of Salt Lake City and 370 miles northwest of Denver.[1] Climate and terrain in the area are harsh and uninviting, but there are large deposits of coal and of trona, a natural form of soda ash, nearby. When the demand for both coal and soda ash began to grow in the early 1970s, Rock Springs became a boomtown. Population, which stood at 11,657 in 1970 soared to 24,000 by 1975. To illustrate the significance

TABLE 6

Population of Green River, Rock Springs, and Sweetwater County,
Wyoming, Selected Years

Year	Green River	Rock Springs	Sweetwater County
1950	3,187	10,857	22,017
1960	3,497	10,371	17,920
1970	4,196	11,657	18,391
1975	9,000	24,900	38,300

Source: Joe Garrett Baker, Labor Allocation in Western Energy Development, Monograph no. 5 (Salt Lake City: University of Utah, Human Resource Institute, 1977), p. 32.

of this change, Table 6 shows the relative stability of the population of this area from 1950 to 1970.

The major energy development project was the Jim Bridger Power Plant, consisting of the sequential construction of four 500-megawatt generating units and an associated coal mine. The Bechtel Corporation was awarded the construction contract for both the generating units and the coal mine. The company surveyed the local situation to determine available labor supply, transportation facilities, social overhead capital, and the structure of governmental units. Based on the architectural and engineering specifications for the plant, a computer model was used to forecast the amount and timing of labor, materials, and capital that were required to do the job. Construction of the first unit started on time. The firm had estimated that about 1,500 workers would be required during the peak construction phase. Figure 3 shows estimated and actual employment on the Jim Bridger project between 1972 and 1977.

As construction of the first generating unit progressed, several things happened. First, three large trona mines in the area began expanding in response to price rises in the soda ash market. Second, two coal mines started construction. Third, as workers and their families moved into the area, construction began on new housing, trailer-park facilities, and roads. The power plant construction company had an increasingly difficult time requiring additional workers as the construction phase accelerated. Lesson number one: the power plant model failed to account for the simultaneous development of other projects requiring relatively large numbers of workers.

FIGURE 3

Jim Bridger Construction Employment

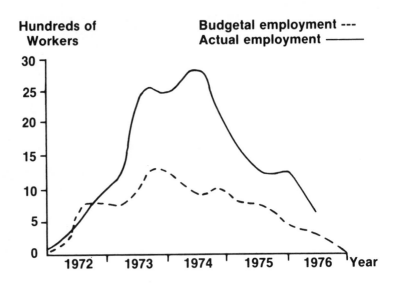

Source: Joe Garrett Baker, Labor Allocation in Western Energy Development, Monograph no. 5 (Salt Lake City: University of Utah, Human Resource Institute, 1977), p. 36.

As the influx of people continued, living conditions in the area began to decline. The lack of adequate housing, schools, hospitals, and other social services became apparent. In addition, rising crime poor roads, absence of recreational outlets, and many related deficiencies placed further hardships on workers and their families. Competition for workers pushed wages up, but the poor living conditions negated many of the monetary gains. Lesson number two: meeting the needs of a rapidly growing population requires advanced planning and the formation of social overhead capital. By and large city and county officials in boomtown areas lack the capability to pla adequately for rapid growth.

As the supply of available workers diminished, the companies were compelled to hire inexperienced workers and pay them a journeyman's rates. This reduced productivity because jobs that previously had called for one skilled worker now required two or more inexperienced employees. The utilization of these workers put additional pressure on the already overextended community infrastructure. Lesson number three: underestimating worker availability may require employing larger numbers of workers, which will, in an impacted community, aggravate the inadequacy of the social infrastructure. While Bechtel Corporation had estimated a peak construction work force of 1,500, employment, in fact (as Figure 3 shows), expanded to 3,000.

As social conditions continued to deteriorate, the psychic costs of staying in the area outweighed the monetary gains. As a consequence, high-wage workers began leaving the area, and absenteeism began to increase among workers still on the job. If the major firms wanted to keep production moving ahead, they had to overhire to ensure a sufficient number of workers. In response to the rising absenteeism, high quitting rates, and low productivity, firms also implemented more aggressive recruitment schemes. The labor force at the Jim Bridger plant organized, and the union attempted to advertise for workers through its national affiliates. The union was able to provide all of the construction workers the company had estimated it would need. However, workers moved out as fast as they moved in. Turnover reached astronomical proportions in all industries. One motel had approximately a 700 percent turnover in 1975, while some of the higher-paying construction and mining firms had rates of over 100 percent.[2] Lesson number four: while workers are fairly mobile and will move to high-wage areas, the social environment must be made acceptable in order to keep them.

The trona-mining firms attempted to recruit long-wall miners from the area, the region, and the entire country, but with little success. Therefore, they were compelled to hire British and Canadian miners. While the company felt "these foreign workers have proved to be good employees . . . this type of recruitment is impractical on a large scale."[3] Lesson number five: foreign workers may be available and may be good employees, but the process of recruitment is costly and probably impractical on a large scale.

It is critically important to recognize that while there are at least two distinct, competing labor markets in the impacted community, the competition is grossly out of balance. The high-wage energy project draws workers away from lower-wage local construction and local service industries. As a consequence, turnover is high and quality of service is low in local industries. Residents of the town who operate or work in one of the lower-wage industries find the increasing cost of living and the deteriorating level of service irritating and divisive. Lesson number six: the lower-wage local industries serve as conduits or staging areas for workers desiring to work in the energy sector.

The cycle of events that caused the severe labor market imbalance in Rock Springs is easy to summarize. The major firms overlooked the simultaneous development of other large projects; the community infrastructure could not support the influx of new residents; the standard of living fell dramatically; skilled workers began leaving the area and firms had to hire less qualified workers; productivity fell; absenteeism made more hiring necessary; and living conditions continued to deteriorate. The attraction of high wages drew workers to the area, but as the noneconomic costs of remaining there exceeded monetary benefits, workers left.

Conclusion

Rock Springs is a larger community than most boomtowns, but the impact of energy development there is not unique. Energy booms affect other communities in a similar manner. In most cases, communities share similar structural characteristics that give rise to the impact problem. These characteristics include:

A relatively small population base,
Slow growth, or in some cases, absolute declines,
A geographically remote location,
An agriculturally based economy (Rock Springs, which had a transportation-based economy, is atypical),
Full utilization of basic infrastructure components (sewers, water, gas), and
No planning capacity within local governmental units.

In terms of outlining what can be done, it is significant that most initiatives must come from within the community.[4] The problem is that most communities either do not see the impending impact or believe that it can be handled with relatively simple measures. To solve or prevent impact problems, it is necessary to recognize the complexity of the impact.

Small communities cannot afford to monitor the development plans of all energy companies that might begin a project in their area. Even if the community were able to identify a company with prospective development plans, the firm might be reluctant to divulge the extent of their development prospects, for proprietary reasons. Unless the community knows where the plant will be located, how large it will be, when development will begin, how fast it will occur, and what modes of transportation will be used, it will be very difficult to plan for impact mitigation.

A final point needs to be made. In developing areas, a major policy involves the participation of indigenous workers in the energy projects. From the standpoint of short-run expediency, many firms would prefer to bring in their construction work force, build the facility, and leave behind the smaller operation and maintenance work force when construction is completed. The problem is that in these situations, segments of the local work force may be prevented from getting a share of the project's higher wages, but will incur the higher prices and other negative impacts that energy development entails. Therefore, it is essential that sufficient planning be done to ensure recruitment and training of local workers. This is particularly important during periods of high unemployment or when development occurs on or near Indian reservations.

NOTES

1. Much of the material in this section is derived from Joe Garrett Baker, Labor Allocation in Western Energy Development, Monograph no. 5 (Salt Lake City: University of Utah, Human Resource Institute, 1977).
2. Ibid., p. 50.
3. Ibid., p. 45.
4. For an excellent summary of how communities can respond to these problems, see Lawrence Susskind and Michael O'Hare, "Managing the Social and Economic Impacts of Energy Development," Summary Report: Phase I of the MIT Energy Impacts Project (Cambridge: Massachusetts Institute of Technology, Laboratory of Architecture and Planning, December 1977).

4
COAL, ELECTRICITY, AND EMPLOYMENT

Regardless of the relative desirability of solar, geothermal, wind, bioconversion, fusion, and other advanced energy technologies, the simple fact remains that they are destined to play a relatively minor role in the nation's energy strategy over the next several years. Consequently, from an employment standpoint, the established energy industries are going to provide the vast majority of employment opportunities through 1985. The Environmentalists for Full Employment are a group with a strong argument suggesting that major employment gains can occur with the initiation of new technologies and expanded conservation programs.[1] We have no basic disagreement with the thrust of this group's position, except in relation to timing. It is our judgment that the commercialization process simply will not permit massive implementation of a solar or a geothermal program during the 1980s. Therefore, their relative employment impact is likely to be rather small.

The big employment programs are likely to be in coal and electricity production. These two industries were expected to produce about 150,000 job opportunities each year between 1977 and 1985. This is broken down into about 25,000 new jobs for miners (both surface and underground) and 125,000 man-years of additional power plant construction work. In the overall scheme of job creation (approximately 2.5 million workers are added to the labor force each year), 150,000 jobs may not seem particularly important. However, the vast majority of these jobs are highly skilled, well-paid positions located within two distinct industrial groups. Employment conditions in both the coal industry and the construction industry are certain to be improved measurably in the near future.

COAL, EMPLOYMENT, PRODUCTIVITY, AND PRODUCTION

The president's energy plan includes the production of 1.2-to-1.3 billion tons of coal each year by 1985. To expand coal production to this magnitude will require production increases of 70-to-80 million tons per year. This works out to a 12-to-15 percent increase each year. Many constraints may impede this expansion; among the most important of these are environmental and land-use restrictions, capital availability, transportation, and labor availability. A critical dimension of the latter factor involves the productivity of the coal-mining work force.

Mine productivity is determined by the type of mining (strip, auger, underground); the character of the coal seams (thick versus thin seams, foreign matter in the coal, and pitch of the coal seam); technology used (continuous miners, conventional techniques, or long-wall methods); application of safety procedures; age and skill level of the work force; length of the work day; and other factors. The precise relationship between these variables changes over time, but there are reasonable grounds to suggest that there is relative stability for short periods.

The trend to lower productivity, in general, started in 1970 and continues today. There is no evidence that there will be a sharp reversal of that trend, because the factors that determine productivity all point toward either stabilized or possibly decreased productivity. Industry does not foresee any substantive improvements in technology; less productive coal seams are being mined; safety standards must become increasingly stringent; and the age and skill level of the work force are declining. In addition, as the 1977/78 coal strike so vividly proved, labor problems are endemic to the industry.

Therefore, a critical question facing the coal industry concerns the change that might occur in miner productivity as rapid production increases are demanded. It will be extremely difficult to reach the president's production goals without productivity increases, and it may approach the impossible if productivity decreases further.

The productivity-production relationships require an examination of how and where increased production will occur, and what can be expected in terms of worker productivity as measured by mining methods and geographic location. Between 1975 and 1985, production capacity is expected to increase by 777.5 million tons per year.[2] Surface capacity should expand by 508.95 million tons per year and underground capacity by 268.55 million tons. On the simplistic assumption that miners work 2,000 hours per year (250 eight-hour days) and that the 1976 productivity estimates apply over the ten-year interval, 79,835 surface miners and 126,376 underground miners will

TABLE 7

Employment: Coal Industry, 1975

Region	Underground	Surface	Total
East of Mississippi River	125,700	45,020	170,720
West of Mississippi River	9,000	10,150	19,150
Total	134,700	55,170	189,870

Sources: Estimates (identical) from the National Coal Association and the Bureau of Mines.

TABLE 8

Employment: Coal Industry, 1976

Region	Underground	Surface	Total
East of Mississippi River	130,602	46,533	177,135
West of Mississippi River	5,375	8,958	14,333
Total	135,977	55,491	191,468

Source: Estimates from the Mining Enforcement and Safety Administration.

TABLE 9

Coal Production, 1976
(thousands of tons)

Region	Underground	Surface	Total
East of Mississippi River	274,356	255,699	530,055
West of Mississippi River	18,470	96,386	114,856
Total	292,826	352,085	644,911

Source: Estimates from the Bureau of Mines, "Coal—Bituminous and Lignite in 1976," Mineral Industry Surveys, table 8, p. 11.

TABLE 10

Projected Increases: Coal Production, 1975-85
(thousands of tons)

Region	Underground	Surface	Total
East of Mississippi River	183,850	49,550	233,400
West of Mississippi River	84,700	459,400	544,100
Total	268,550	508,950	777,500

Source: George F. Nielsen, "Coal Mine Development and Expansion Survey . . . 617.3 Million Tons of New Capacity 1977 through 1985," Coal Age, February 1977, pp. 84-91.

TABLE 11

Estimated Coal Production, 1985
(thousands of tons)

Region	Underground	Surface	Total
East of Mississippi River	458,205	305,249	763,455
West of Mississippi River	103,170	555,786	658,956
Total	561,375	861,035	1,422,411

Source: Combination of Tables 9 and 10.

TABLE 12

Increase in Employment: Coal Production, by Region, 1985

Region	Underground	Surface	Total
East of Mississippi River	86,518	7,773	94,291
West of Mississippi River	39,858	72,062	111,920
Total	126,376	79,835	206,211

Source: Based on Table 11.

TABLE 13

Total Employment: Coal Production, by Region, 1985

Region	Underground	Surface	Total
East of Mississippi River	212,218	52,793	265,011
West of Mississippi River	48,858	82,212	121,070
Total	261,076	135,005	386,081

Source: Combination of Tables 7 and 12.

TABLE 14

Employment: Percentage Change, 1976-85

Region	Underground	Surface	Total
East of Mississippi River	69	17	55
West of Mississippi River	443	710	584
Total	94	145	109

Source: Combination of Tables 7 and 12.

be needed ($250 \times 8.5 = 2,125$; $250 \times 25.5 = 6,375$; $508,950,000 \div 6,375$ = 79,835; $268,550,000 \div 2,125 = 126,376$). This represents a doubling of the number of miners nationwide. It is extremely important to point out that the number of miners will approximately double if the pattern of projected production occurs. However, if the best available control technology (scrubbers) were mandated for all power plants, then coal from the East would be competitive with western low- sulfur coal. Under these conditions, the number of miners could triple.

More important, about 68.5 percent of the underground capacity increases are scheduled east of the Mississippi River, while only 9.7 percent of the surface capacity will develop in this region. In 1975, the distribution of miners was as shown in Table 7. In 1976, data produced by the Mining Enforcement and Safety Administration (MESA) showed the distribution given in Table 8.

Estimated production for 1976 is shown in Table 9. For 1985, the projected increase in production is that shown in Table 10. Though the expansion of capacity in 1976 does not equate with projected increases in production for 1975-85, as a proxy for the expanded pattern of production, the combination of Tables 9 and 10 produces the data shown in Table 11.

Based on Table 11, and the simplified assumptions noted above, a tentative geographic distribution of the employment impact can be developed, as shown in Table 12. Total employment in each area in 1985 is estimated in Table 13. The relative increases in employment by region, between 1976 and 1985, are shown in Table 14.

A 10 percent increase in the total mining work force per year does not appear difficult to achieve, but a 60 percent projected annual increase in the western region of the country is a more serious matter. When certain geographic and population characteristics of the western states are considered, such as the remote location of many mine sites and the small population base, the difficulty of increasing the mining work force there becomes even more apparent.

Comparisons by state produce essentially the same picture. The 15 largest capacity increases by state are shown in Table 15. In addition, the 1975 productivity estimates are shown for both surface and underground mining. Only West Virginia (southern), Pennsylvania, Alabama, and Ohio, among the largest-capacity states, have productivity estimates for underground mining that are below the national average of 8.5 tons per day. Five states, West Virginia (northern and southern), Kentucky (eastern), Illinois, Pennsylvania, and Alabama have surface-productivity estimates below the national average of 25.5 tons per day.

The four states in the below-average underground-mining group will increase capacity by about 99 million tons per year by 1985 (about 36 percent of the national total of 268.55 million tons). The six below-

TABLE 15

Production–Productivity in Major Coal–Producing States

State	Production Capacity Increase (million tons)			Productivity (tons per day)	
	Total	Underground	Surface	Underground	Surface
Wyoming	220.50	4.00	216.50	10.79	67.74
Montana	75.10	–	75.10	–	127.25
Utah	59.20	46.70	12.50	13.85	n.a.
North Dakota	57.90	4.00	53.90	n.a.	86.86
West Virginia (southern)	49.95	47.55	2.40	8.24	17.01
Colorado	49.25	27.00	22.25	10.43	45.62
Kentucky	41.55	35.65	5.90	11.36	22.30
Texas	40.60	–	40.60	–	76.49
Illinois	38.40	24.70	13.70	14.25	24.19
Pennsylvania	26.15	23.15	3.00	8.23	20.59
New Mexico	25.40	–	25.40	–	60.44
Alabama	1.50	17.35	1.50	7.23	15.29
Ohio	3.10	10.90	3.10	8.19	25.97
West Virginia (northern)	5.50	8.20	5.50	11.34	20.20
Indiana	11.70	0.50	11.70	16.10	29.69

n.a. = not available

Sources: Bureau of Mines (productivity), "Coal—Bituminous and Lignite in 1975," Mineral Industry Surveys, February 19, 1977, table 15, p. 18; George F. Nielsen (production), "Coal Mine Development and Expansion Survey . . . 617.3 Million Tons of New Capacity 1977 through 1985," Coal Age, February 1977, pp. 84–91.

TABLE 16

Productivity, 1972-76
(tons per day)

State	1976*	1975	1974	1973	1972
Alabama	8.96	11.19	12.49	15.79	16.42
Alaska	40.14	30.65	33.36	33.86	25.96
Arizona	65.75	69.66	78.95	76.90	130.35
Arkansas	8.31	8.25	11.34	12.35	12.74
Colorado	16.12	18.89	19.56	17.46	16.83
Georgia	7.72	9.79	—	—	—
Illinois	16.79	17.61	20.03	23.56	24.39
Indiana	27.87	29.50	32.02	35.98	33.75
Iowa	20.19	20.15	20.46	21.20	22.04
Kansas	20.02	13.76	13.51	16.20	17.07
Kentucky					
Eastern	—	15.40	16.84	18.13	17.71
Western	—	20.22	24.14	28.02	28.89
Total	13.90	16.99	19.26	21.29	21.26
Maryland	16.84	20.69	24.54	28.20	26.55
Missouri	27.35	21.14	20.02	30.39	29.14
Montana					
Bituminous	—	129.66	134.42	132.25	139.80
Lignite	—	53.59	59.28	55.55	63.94
Total	115.23	127.25	130.94	127.11	133.60
New Mexico	41.24	36.86	46.66	48.84	54.92
North Dakota	101.85	86.86	66.83	102.36	101.88
Ohio	12.35	15.13	17.01	21.26	22.59
Oklahoma	14.15	14.79	17.66	19.82	19.43
Pennsylvania	11.22	11.46	13.68	12.52	12.82
Tennessee	16.19	12.94	15.90	20.54	20.41
Texas	74.33	76.49	80.50	100.75	55.22
Utah	11.96	13.85	15.32	14.36	13.49
Virginia	8.32	10.69	11.88	12.65	12.62
Washington	40.22	27.11	44.31	38.90	49.44
West Virginia	7.63	9.15	10.79	11.83	12.55
Wyoming	37.78	61.78	65.93	55.94	52.01
Total	—	14.74	18.68	17.58	17.74

*1976 data are based on MESA estimate using 250 work days. For comparative purposes, use 1972-75 data.

Source: Bureau of Mines, "Coal—Bituminous and Lignite," Mineral Industry Surveys, table 10, p. 340.

average surface-mining states will expand capacity by about 32 million tons per year (about 6 percent of the national total of 508.95 million tons). In both instances, the low-productivity states are east of the Mississippi River. Factors such as outdated technology and less productive coal seams probably account for most of the productivity differential.

The pattern of productivity changes for 1972-75 is shown in Table 16. From 1972 to 1975, average productivity declined from 17.74 tons per day to 14.74 tons per day (a decrease of about 17 percent). * Average productivity in four of the top ten producing states increased during this period as follows:

Rank	State	Increase in Productivity (tons per day)	Percent Increase
6	Colorado	16.83-18.89	12
8.	Texas	55.22-76.49	39
3	Utah	13.49-13.85	3
1	Wyoming	52.01-61.78	19

The remaining six producers among the top ten had productivity decreases as follows:

Rank	State	Decrease in Productivity (tons per day)	Percent Decrease
9	Illinois	24.39-17.61	28
7	Kentucky (eastern)	17.71-15.40	13
2	Montana	133.60-127.25	5
4	North Dakota	101.88-86.86	15
10	Pennsylvania	12.82-11.46	11
5	West Virginia	12.55-9.15	27

Only two states, Illinois and West Virginia, had productivity declines exceeding the national average.

There seem to be reasonable grounds to argue that the high-production states correspond in some sense to the high-productivity states. While there is some volatility in the individual state productivity estimates, in most instances the trend is clearly downward.

Without substantive advances in technology, employment in the coal industry will be very sensitive to declines in productivity. If

---

*The 1976 estimates should be discounted because they are preliminary estimates based on MESA data. In addition, the 1976 estimates assume a standard 250-day work year.

production objectives are set at X tons per year and it requires Y workers to achieve this goal, a 10 percent reduction in the productivity of the work force will prompt about a 10 percent increase in the size of the work force. Costs of production rise and are translated into increased coal prices. It can be argued that price increases will slow coal demand, but government regulation and existing technology should make the short-run demand for coal relatively price inelastic.

ELECTRICITY AND JOBS

In the energy sector the single largest producer of new employment opportunities will be the construction of electric power plants. The Department of Labor's Construction Labor Demand System (CLDS) staff has estimated that about 64,000 man-years of employment will occur each year between 1977 and 1981 in the construction of nuclear power plants; 54,000 in the building of fossil-fueled (mostly coal) power plants, and about 7,000 in hydroelectric plants. *

The data are disaggregated into 29 occupational classifications and are arrayed by month, for various geographic areas. Analysts interested in obtaining estimates for their particular area may, of course, contact the Department of Labor. These estimates are updated monthly, and therefore for planning purposes, there is a continual stream of information that cannot be readily captured in this study. Nevertheless, it may be useful to summarize a recent data run to suggest the character of the available data.

Power Plant Construction
in the United States, 1977-81

In the aggregate, between 1977 and 1981, there will be a total of 623,589 man-years of work in the construction of nuclear, fossil-fueled, and hydroelectric power plants. The estimates, by plant type, are shown in Table 17.

There is no convenient method of determining the precise number of jobs a specific number of man-hours will generate in the con-

---

*The CLDS, a system for estimating the manpower requirements in the energy sector, is described fully in Chapter 7. It was originally called the Construction Manpower Development System (CMDS), but the name was changed to CLDS in 1978. We will refer to the system as the CLDS, except in citing some pre-1978 sources that refer to it as CMDS.

TABLE 17

Estimated Man-Years in the Construction of Nuclear, Fossil-
Fueled, and Hydroelectric Power Plants, 1977-81

| Man-Years | Plant Type | | | |
	Nuclear	Fossil-Fuel	Hydro	Total
1977	52,253	50,335	7,278	109,866
1978	61,360	55,186	7,390	123,936
1979	68,310	55,400	6,527	130,237
1980	70,375	51,737	7,399	129,511
1981	67,535	56,239	6,265	130,039
Total	319,833	268,897	34,859	623,589

Source: U.S., Department of Labor, Construction Labor Demand System.

TABLE 18

Estimated Employment in the Construction of Nuclear, Fossil-
Fueled, and Hydroelectric Power Plants, 1977-81

| Employment | Plant Type | | | |
	Nuclear	Fossil-Fuel	Hydro	Total
1977	60,091	57,885	8,370	126,346
1978	70,564	63,464	8,499	142,527
1979	78,557	63,710	7,506	149,773
1980	80,931	59,498	8,509	148,938
1981	77,665	64,675	7,205	149,545
Total	367,808	309,232	40,089	717,129

Source: U.S., Department of Labor, Construction Labor Demand System (jobs/man-year conversion factor equal to 1.15).

TABLE 19

Man-Hours of Employment in Nuclear, Fossil-Fueled, and Hydroelectric Power Plants, by Occupation, 1977-81

Craft	1977	1978	1979	1980	1981	Total (all years)
Pipefitters	31,221.1	40,569.4	41,902.4	42,166.4	42,747.1	198,606.4
Laborers	33,119.1	35,960.0	35,326.5	35,233.4	33,816.7	173,455.7
Electricians	25,685.8	31,753.6	36,016.0	35,215.3	36,043.4	164,714.1
Carpenters	24,639.0	25,831.6	24,967.7	24,279.0	23,919.7	123,637.0
Operating engineers	17,853.1	19,122.3	18,346.6	17,906.9	17,354.1	90,583.0
Pipefitter-welders (nuclear certified)	9,414.5	11,570.8	13,051.0	13,099.9	13,256.1	60,392.3
Iron workers—structural	10,399.0	11,621.5	11,646.8	11,426.7	11,667.1	56,761.1
Boilermakers	11,643.8	13,776.2	13,675.1	13,190.7	13,446.8	65,732.6
Iron workers—reinforcing	9,311.5	9,368.2	8,842.3	8,471.6	9,003.3	44,996.9
Truck drivers and warehousemen	8,748.6	9,217.9	8,769.6	8,285.8	7,778.5	42,800.4
Boiler welder (nuclear certified)	5,792.7	6,859.5	6,823.2	6,592.1	6,683.8	32,751.3
Millwrights	4,859.4	5,792.1	6,295.0	6,224.2	6,373.1	29,543.8
Asbestos workers	97.8	4,539.6	5,565.2	5,583.3	5,726.9	21,512.8
Painters	3,640.4	4,122.8	4,621.3	5,015.1	5,476.5	22,876.1
Sheet-metal workers	2,892.0	3,443.4	4,067.3	3,843.1	4,164.8	18,410.6
Cement masons	2,613.8	2,862.9	2,866.5	2,769.1	2,824.8	13,937.1
Bricklayers and Stonemasons	896.1	1,009.3	988.9	1,026.1	1,039.9	4,960.3
Other construction craftsmen	871.7	917.2	914.5	870.0	844.8	4,418.2
Total	203,699.4	238,338.3	294,685.9	241,198.7	242,167.4	1,164,064.6

Source: U.S., Department of Labor, Construction Labor Demand System.

44

struction industry. However, the man-year estimates can be converted into jobs estimates by using a conversion factor of about 1.15, as shown in Table 18. This is only an approximation, however, and caution must be used in interpretation. The conversion factor reflects a number of years of experience in measuring the jobs/man-year relationship. Factors such as absenteeism, lateness, vacations, and other forms of reduced work time require slightly more than one worker for each man-year of effort required.

Since these data reflect power plant construction occurring in a variety of phases, the consistency of the estimates is striking. For 1978-81, the estimate for total jobs varied from 142,527 to 149,773, which is about a 5 percent variation. Preliminary data gathered by the CLDS staff suggest that there will be a sharp increase in total jobs by the mid-1980s.

The CLDS data can be disaggregated into 29 occupational areas, as noted above. Some of these areas do not appear in power plant construction, however, and therefore will be omitted from the following analysis. These areas include lathers, plasterers, glaziers, and several others. Table 19 shows the occupational distribution for the nation.

Over the 1977-81 period, pipefitters, laborers, electricians, and carpenters will have the largest number of employment opportunities, all exceeding 6,400 man-years of effort. The two welder categories, boilermaker-welder and pipefitter-welder, require welders who are nuclear certified.

Pipefitters comprised 15.0 (31,221.1 ÷ 207,700.5 = 15.0) percent of the on-site work force in 1977, but will expand to 17.7 (42,747.1 ÷ 242,139.3 = 17.7) percent in 1981. The second largest skilled occupation is electrician. The electricians' share of the on-site labor force in 1977 was 12.4 (25,685.8 ÷ 207,700.5 = 12.4) percent; in 1981 it will rise to about 14.9 (36,043.4 ÷ 242,139.3 = 14.9) percent. Carpenters and laborers will both experience relative declines from 11.9 (24,639.0 ÷ 207,700.5 = 11.9) percent and 15.9 (33,119.1 ÷ 207,700.5 = 15.9) percent, respectively, in 1977 to 9.9 (23,919.7 ÷ 242,139.3 = 9.9) percent and 14.0 (33,816.7 ÷ 242,139.3 = 14.0) percent, respectively, in 1981.

Using 1977 as the base year, several comparisons showing relative changes in demand, by occupation, are of interest. Overall, there will be a 16.6 percent increase over the 1977 level. The relative annual changes are:

Year	Percent Change
1977/78	13.3
1978/79	4.0
1979/80	-1.5
1980/81	0.4
1977-81	16.6
1977-peak	17.8

For the last part of the 1970s, construction-worker demand in the energy sector apparently flattened out. However, as noted above, preliminary information suggests that there will be a surge of activity after 1981.

Regional Profiles

There is a definite regional pattern of power plant construction. About 57.4 percent of the construction-job opportunities between 1977 and 1981 will occur in regions 4, 5, and 6, listed in Table 20. Two of these regions would be classified as southern regions. The population of these regions for 1971 and projections to 1980 and 1990 are shown in Table 20.

In 1971, about 47.4 percent of the population lived in the 19 states included in these regions. By 1980 this percentage will have increased to about 47.8 percent; by 1990 it will be about 48.1 percent. The projected change in the nation's population between 1971 and 1980 was 8.4 percent. Between 1971 and 1990 the population will increase by about 19.3 percent. Both regions 5 and 6 will experience growth rates that are slower than the national average. Nevertheless, over 36 percent of the new power plant construction will occur in these two regions. Inferentially, it may be argued that new power plant construction is not following the predicted population shifts.

However, even though regions 5 and 6 will not experience substantial population growth, they are attractive locations for manufacturing. The outlook for manufacturing calls for a doubling in earnings across the nation between 1971 and 1990, but predictions for regions 5 and 6 show increases significantly above the 100 percent figure. This anticipated growth in manufacturing could be the cause of the demand for more power plants. It may seem plausible to argue that per capita electricity consumption will rise in the Sunbelt areas of regions 4 and 5 as the industrialization process gains momentum. However, per capita income in these regions will continue to be somewhat lower than the national average, so the increased demand points to a combination of factors heavily dependent on increased manufacturing.

TABLE 20

Population by State, 1971-90

State	1971	1980	1990	Percent Change, 1971-80	Percent Change, 1971-90
Region 4					
Alabama	3,487,000	3,487,000	4,090,000		
Florida	7,026,000	8,926,400	10,978,100		
Georgia	4,646,180	5,147,300	5,907,400		
Kentucky	3,276,000	3,608,800	3,982,300		
Mississippi	2,250,000	2,327,900	2,450,300		
North Carolina	5,158,000	5,736,300	6,464,700		
South Carolina	2,633,000	2,818,500	3,121,900		
Tennessee	3,994,000	4,556,800	5,190,500		
Total	32,470,180	36,609,000	42,185,200	13.5	29.9
Region 5					
Illinois	11,182,000	12,090,900	13,056,400		
Indiana	5,244,100	5,783,600	6,364,200		
Michigan	8,996,000	9,742,500	10,645,100		
Minnesota	3,860,000	4,119,400	4,553,100		
Ohio	10,739,000	11,650,600	12,609,400		
Wisconsin	4,473,000	4,736,600	5,012,900		
Total	44,494,100	48,123,600	52,241,100	8.2	17.4
Region 6					
Arkansas	1,951,000	2,086,600	2,271,000		
Louisiana	3,693,100	3,744,300	3,936,500		
New Mexico	1,045,000	1,054,900	1,131,200		
Oklahoma	2,600,000	2,762,300	2,993,400		
Texas	11,428,000	12,166,900	13,579,700		
Total	20,717,100	21,815,000	23,911,800	5.3	15.4
U.S. total	206,188,000	223,532,000	246,039,000	8.4	19.3

Source: U.S., Department of Commerce, Area Economic Projections, 1990 (Washington, D.C.: Government Printing Office, 1974).

TABLE 21

Man-Years of Employment in Power Plant Construction, by Region, 1977-81

Region	1977	1978	1979	1980	1981	Total
1	2,461.2	3,668.4	5,302.8	6,877.2	6,645.6	24,955.2
2	4,960.8	5,126.7	5,810.9	7,723.6	9,146.2	32,768.2
3	10,363.2	11,080.0	9,922.5	7,543.2	7,818.8	46,727.7
4	23,461.2	22,541.4	23,930.3	25,977.4	30,551.1	126,461.4
5	25,596.0	25,052.5	26,055.8	25,073.6	22,777.1	124,555.0
6	17,822.0	19,153.8	19,846.8	17,749.3	15,896.3	90,468.2
7	8,696.4	9,685.5	11,086.3	9,585.3	7,143.9	46,197.4
8	6,360.0	6,474.6	6,232.6	5,806.6	6,118.2	30,992.0
9	7,801.2	7,980.5	8,742.5	8,947.1	9,917.5	43,388.8
10	4,563.6	5,366.2	6,338.4	6,276.8	6,271.8	28,816.8
Total	112,085.6	116,129.6	123,268.9	121,561.1	122,286.5	595,315.9

Note: These data are 4.7 percent below the national estimate (623,589) because 25 plants representing 3,621 megawatts are unsited. In brief, unsited means that a company has indicated that additional capacity will be built, but it has not decided definitively on a site. Hence, it is in the national totals, but not the regional data.

Source: U.S., Department of Labor, Construction Labor Demand System.

TABLE 22

Changes in Man-Year Requirements for Power Plant Construction,
by Region, 1977-81

Region	1977/78	1978/79	1979/80	1980/81	1977-81
1	49.1	44.5	29.7	-3.4	170.0
2	3.6	13.3	33.6	17.8	84.9
3	6.9	-10.4	-24.0	3.7	-24.5
4	-3.9	6.2	8.6	17.6	30.2
5	-2.1	4.0	-3.8	-9.2	-11.0
6	7.5	3.6	-10.6	-10.4	-10.8
7	11.4	14.5	-13.5	-25.5	-17.8
8	1.8	-3.7	-6.8	5.4	-3.8
9	2.3	9.5	2.3	10.8	27.1
10	17.6	18.1	1.0	0.1	37.4
Nation	13.3	4.0	-1.4	0.4	16.1

Source: U.S., Department of Labor, Construction Labor Demand System.

Regions 1 through 5 are east of the Mississippi River and 6
through 10 are west of the Mississippi River. Table 21 shows that
about 40 percent of power plant construction will occur in the five
western regions. However, the pattern of expansion differs markedly.
There is a continual upward movement from about 67,000 man-years
in 1977 to 77,000 man-years in 1981 in the eastern regions. In the
western regions the man-years of employment increase until 1979 and
then decrease during the subsequent two years. In the aggregate,
this pattern of development may suggest that after 1979 efforts should
be made to encourage surplus workers in the West to move to the East.
 While detailed breakdowns are available for the ten regions,
summary estimates for man-year requirements for each region suggest where the maximum employment potential exists. Table 22 shows
that the man-year requirements will increase about 16.1 percent between 1977 and 1981. Five regions will experience net relative increases between 1977 and 1981 and five will experience losses. Specific crafts will show even greater relative changes. For example,
painters in region 1 will show an increase of from 43.2 man-years of

employment in 1977 to 363.6 man-years in 1981 (a 741.7 percent increase). In region 7, boilermakers and boilermaker-welders will experience 55.8 percent and 55.7 percent decreases, respectively, over the 1977-81 interval. The analysis of specific craft data, by different geographic areas, will provide planners with important information about where shortages are likely to occur.

CONCLUSION

The rapid expansion of electricity production and the concurrent development of new coal mines will provide most of the new employment in the primary energy industries through 1981. The other industries comprising the energy sector will also increase, but their net changes are not expected to exceed normal long-term growth patterns. There may be increased expansion in railroad transportation, gas extraction, and probably construction and mining equipment. In fact, the Bureau of Labor Statistics (BLS) estimates that the following increases are likely to occur:

Industry	1976	1980	1985
	(in thousands)		
Railroad equipment	43	53	57
Oil- and gas-well drilling and exploration	195	250	234
Construction, mining, and oilfield machinery	248	268	311

The employment-creating potential of the entire energy sector between 1977 and 1985 is estimated by the BLS to increase by about 16 percent (translating into about 60,000 new jobs each year). The estimated 823,589 man-years of work in power plant construction, and the extrapolated increase of 103,000 new jobs in the coal industry between 1977 and 1981, show that these two industries will carry the bulk of new energy-sector employment through 1981 (and quite likely to 1985 and beyond).

The generation of electric power in the United States involves an established technology that is unlikely to change materially during the next several decades. The employment requirements for the construction and operation of these facilities are well established, so we can determine quite precisely the number of workers required, the skills mix, and timing of labor input for electricity-generation projects. The labor requirements, particularly in terms of skills mix, do vary among fossil-fueled, nuclear, and hydroelectric plants in both construction and operation. Therefore, the unresolved question is: What mix of generation facilities will emerge over the next several decades? It is reasonable to argue that major employment-estimation problems will not develop in this energy industry.

On the other hand, other energy sources are likely to pose serious human resource planning problems. The United States must develop alternative sources of liquid and gaseous fuel and presumably exploit geothermal resources and solar energy. There are a variety of competing but essentially unproved technologies related to liquid and gaseous fossil-fuel production. Most important among these experimental technologies are oil shale, tar sands, bioconversion, coal liqeufaction and gasification, and exploitation of deep geopressured, geothermal methane resources. There are even more esoteric schemes such as utilizing ocean thermal gradients, wind, and wave motion.

Since we have very limited experience with these technologies, the human resource requirements for each are essentially unknown. Even a cursory search reveals that there are no definitive studies related to solar, geothermal, or several of the more advanced technologies, such as oil shale and coal gasification. Clearly, one of the major challenges for energy/employment analysts over the next several decades will be to track these emerging technologies and determine their relative employment impacts.

In addition to the emerging technologies themselves, it should be noted that there has been considerable discussion about one element of the overall alternative-technology program. The essence of this approach is that major conservation, linked with small-scale, individualized energy systems, is a viable (some say preferable) alternative to large-scale energy systems. The proponents of this view argue that it will create employment.[3] However, to date their analytical efforts have been largely anecdotal. Nevertheless, if a small-scale energy-systems approach should emerge as a major segment of our long-run energy strategy, the employment consequences could be dramatic.

NOTES

1. Environmentalists for Full Employment, Jobs and Energy (Washington, D.C.: Environmentalists for Full Employment, Spring 1977).

2. George F. Nielson, "Coal Mine Development and Expansion Survey . . . 617.3 Million Tons of New Capacity 1977 through 1985," Coal Age, February 1977, pp. 83-100.

3. Environmentalists for Full Employment, op. cit.

5
ENERGY AS A FACTOR
IN PRODUCTION/
CONSUMPTION PROCESSES

The relationship between the growth in energy use and the growth in GNP and employment is one of the most critical policy questions of our day. Both the availability and price of energy have a great impact on the location of industry and employment. What role will energy play in determining the growth in total employment? How will it affect the relative development of metropolitan areas? How will the warm Sunbelt states develop in relation to the rest of the nation? Will high prices curtail expansion of energy-intensive industry in favor of industry that requires relatively small quantities of energy for given levels of output?

Raising this set of questions is not to suggest that answers will be provided in this or any other study in the near future. It is imperative that analysts improve their ability to assess the relationships between energy sources and uses and changes in employment, by industry, occupation, and area. Development of these assessment tools will aid decision makers in their understanding of the impacts of various policy options on employment. Presently, it is very difficult to forecast the employment aspects of alternative energy policies. Some energy options can be viewed as having both job-creating and job-destroying potential, although it is difficult to determine short- or long-run net impacts on employment. For example, a component of the 1977 National Energy Plan (NEP) involves heavy emphasis on conservation. If we are able to structure and sustain a strong conservation program, the impact on employment will be substantial. A policy promoting mass transit and discouraging dependence on the private automobile will also have a far-reaching effect. However, we do not know how fast employment might grow or decrease in relation to energy-policy decisions, or how skills might transfer between alternative strategies.

The complexity of energy issues suggests that it will be necessary to sharpen our measurements of substitutability and complementarity of factors of energy production, the related income effects, and the economic-management analysis of technological change and of changes in production methods. Equally important from the viewpoint of employment policy is the necessity to gain some perspective on the timing and magnitude of market or administrative responses to the energy problem. Planning effective human resource policies requires some sense of the direction, timing, and magnitude of energy policies. An additional issue is the extent to which decisions are guided by income-redistribution goals that protect the poor or by price controls that guard against windfall profits. The multiple variables that affect energy prices make forecasting exceedingly difficult.

As of this writing, contrary to popular belief, energy prices have slowed in their upward spiral. This may be a temporary phenomenon, but it is nonetheless true, not only for gasoline but also for electricity. The major price increases occurred between 1973 and 1975, and prices stabilized during 1976 and 1977. For example, prices for 500 kilowatt hours of electricity rose by 43 percent between 1973 and 1975, but increased only 16 percent between 1975 and 1977. In constant dollars, the increase was 17 percent for the first two-year period and 3 percent for the latter.

Unfortunately, because of a circular chain of interaction, we cannot determine what the behavior of energy prices will be during the next five to ten years. Both the sources and uses of energy are tied to prices. Since energy prices are determined to a large extent by fiat, it becomes very difficult to predict their behavior. Will the Organization of Petroleum Exporting Countries (OPEC) increase prices of crude oil at its next meeting and, if so, by how much? Will the U.S. Congress decontrol gas and oil prices and, if so, on what timetable? Will other western governors follow Montana's lead and impose heavy severance taxes on their coal, oil, or gas?

Energy optimists point out that at the present time, there appears to be both an adequate supply of coal within the country and a ready supply of oil on the international market. There should not be a major shortfall in oil availability (barring another OPEC embargo) until well into the next decade. As inevitable as the ultimate shortage is, the United States—and for that matter, the world at large—has returned to a business-as-usual attitude. However, sooner or later, prices will rise, forcing development of alternative energy sources and more careful use of traditional supplies.

Since there is no precise method of predicting future price activity, we recommend the development of several scenarios that will bracket the probable price changes. Most important will be the analysis of the sensitivity of employment changes to various price changes.

This approach, while clearly inadequate, should provide policy makers with outside parameters of probable employment needs. Unless we go through the exercise of formulating several alternative strategies, it is unlikely that reasoned judgments can be made. No one knows what the employment impact of a 10 percent rise in OPEC prices will be, or how deregulation of gas and oil prices will affect employment. These are important questions that deserve more than anecdotal judgments about likely changes.

There are several macroeconomic issues that seem to have captured the interest of analysts and policy makers. To suggest what the character of the energy/employment issue is, this chapter outlines the current thinking on four of them: the substitutability of labor for energy, the geographic relocation of major industries, marginal analysis and large-scale development projects, and the National Energy Plan's policies related to income redistribution and windfall profits.

THE SUBSTITUTABILITY OF LABOR FOR ENERGY

A major area of inquiry that should be explored in more depth is the potential for substituting labor for energy or capital. As the prices of these latter two factors rise, labor becomes relatively more attractive as a factor of production.

Economists have long postulated that firms typically attempt to find the combination of factor inputs that minimizes the per-unit cost of production. The traditional view has involved the substitution of labor for capital or land at the marginal level, so that the following equality holds:

$$\frac{MP_L}{P_L} = \frac{MP_K}{P_K} = \frac{MP_1}{P_1}$$

where L is labor, K is capital, and 1 is land. If there are any ratios that upset the equality, it will be advantageous to substitute one factor for another until the ratios are again equal. It is easy to show that as long as an inequality exists, it will be possible to produce a specified level of output at a lower level of resource usage.

When energy is considered as a major factor of production, then the question of the degree of substitutability among energy, labor, capital, and land arises. When the price of one factor increases relative to the others, it is to a firm's advantage to substitute the less expensive factor for the more expensive one. Bruce Hannon, of the University of Illinois's Energy Research Group, has suggested that "those inputs with relatively high marginal productivities are used in a relatively profligate way."[1] In any case, even though labor costs have been rising during the last decade, it appears that energy prices have been rising at an even faster rate. As a consequence, it seems logical to argue that firms will, if possible, substitute the lower-priced labor resource for the relatively higher-priced energy resource. Of course, there are technological and practical constraints on factor substitution, but some plausible arguments supporting this hypothesis emerge from a study of energy/employment trade-offs.

Few positive things can be said about the impact of high energy prices. However, Dale Jorgenson has suggested that higher prices could have a positive effect on employment. From a policy standpoint, the critical dimension of the energy/employment relationship involves the degree to which labor and energy are in fact substitutable. If they are only marginally substitutable, then the overall employment effect of higher energy prices may be minimal. Ernst Berndt and David Wood examined the substitutability question in 1975 and concluded that "technological possibilities for substitution between energy and non-energy input are present, but to a somewhat limited extent."[2] More specifically, they found that "energy and labor are slightly substitutable . . . but energy and capital are complementary."[3] This analysis suggests that the overall employment effect of higher energy prices may be relatively small.

Another dimension of the employment effect, proposed by Ron Kutscher of the Bureau of Labor Statistics, is that it may be plausible to argue that skilled, professional, and technical occupations are complementary with energy usage, while unskilled occupations are substitutes for it. If, in fact, these relationships exist, the employment effect of higher energy prices may simply be to change the occupational mix. The nature and dimensions of this change would, of course, be an important finding.

A somewhat different view, but one that still essentially supports the contention that higher energy prices create jobs, was offered by Hannon. He argued that "under conditions of zero economic growth, the United States could have accomplished full employment in the 1935 to 1970 period by raising the price of energy relative to wages."[4] This suggests, of course, that there is a degree of substitutability between energy and labor. Hannon noted that in the past, the price of labor has risen much faster than the price of electricity, and as a consequence, electricity has been substituted for labor. Economic growth helped absorb new workers and prevented a rapid rise in unemployment. But even if economic growth is curtailed, full employment could still be achieved. Hannon further argued, without substantiation, that the rapid drop in the wage/electricity-price ratio between 1970 and 1975 supports the hypothesis that energy and labor are substitutable.[5]

In 1974, over 900,000 new jobs were created for each quadrillion Btu's of energy saved. However, not all energy-saving activities lead to an equal increase in the number of jobs. For example, if Americans change from intercity-plane to intercity-train transportation, about 930,000 new jobs will be created for each quad of energy saved. On the other hand, a similar one-quad saving made by a switch from electric to gas stoves would produce only 160,000 new jobs. Hannon analyzed a variety of energy-saving activities and suggested that "given that the present U.S. energy use is about 80 quads and reducible unemployment at most 4 million persons, full employment could be reached (leaving about 3 percent frictional unemployment) with energy use reduced by approximately 5 to 10 percent."[6] Much of this energy saving could come about by shifting from intercity-plane to intercity-train transportation.

Unfortunately, from an employment perspective, there are some activities that result in decreases in jobs as energy expenditures increase. The most significant in Hannon's view is the increase in electricity generation. He noted that "approximately 75,000 jobs are lost over the entire economy, in the short run at least, for each new quad of primary energy transformed into electricity."[7] The most important component of the electricity/employment relationship is the presumed income effect of increased electricity generation. Hannon believes that jobs are lost because "the decision to purchase electricity requires a reduction in spending somewhere else. Although this spending reduction means a reduction in energy demand, it also means a reduction in the demand for labor; and this reduction exceeds the number of jobs involved, directly or indirectly, in the purchase of electricity."[8]

We are convinced that the income effects of increased energy prices may in the long run be much more important than is generally believed. How these income effects will translate into changes in the mix and level of employment is not known. Some preliminary work at the Center for Advanced Computation at the University of Illinois suggests that the income effects dwarf the substitution effects.[9] The analysis of the 23 percent increase in energy consumption between 1963 and 1967 shows that expanded income accounted for over three-fourths of the increase in usage. Population increases, technological change, and shifts in the market basket of consumer goods determined the remainder of the increase. If expanded real income has played such an important role in the upsurge of energy use, there is some reason to expect that as real consumable income falls, there will be corresponding impacts.

As per capita income increases, people consume more of the same types of goods. These goods require both energy and labor to produce, transport, and consume. The problem is that we do not know what kinds or what amounts of labor are needed. Similarly, as income falls as the result of higher energy prices, a job-loss effect can be inferred. However, the character of this effect is essentially unknown.

RELOCATION OF U.S. INDUSTRY

The shift of U.S. industry from the northeast to the Sunbelt and the Rocky Mountain region is an established fact that requires no documentation. This movement has been prompted by many factors, including shifts in the product or resource market, labor force availability, state legal restrictions, energy availability, and congestion. It might be reasonable to insert the word expected, or anticipated, before each of these factors. Industry expects the resource and the product market to move as the population migrates south and west. Similarly, the labor force will be located where the population is located.

This shift in the location of U.S. industry creates two problems, one on each end of the process. As firms leave an area, its economic base is diminished. Job losses occur, incomes fall, the tax base decreases, and related or support industries consider moving. On the receiving end, these changes are likely to occur in reverse. However, there may be problems on the receiving end as well. If the industry is clean (usually meaning nonpolluting), it is usually welcome in any community. But if it is not clean, or if it is too large in relation to the recipient community, major adjustment problems, such as those discussed in Chapter 3, frequently arise.

There is a need to facilitate labor market functions in both declining and growing regions. The 1974 Trade Act provides resources and retraining or relocation assistance to those workers who have lost their jobs due to our nation's international trade policy. It seems logical to argue that similar assistance should be available to workers who, through no fault of their own, lose their jobs when industry relocates and leaves them without employment. Unemployment insurance can assist these workers; however, these benefits are limited in size and duration, and provide essentially no long-run solution to the problem of workers displaced by relocation. In a declining community jobs are scarce or nonexistent. Unless some form of external aid is offered, those left behind will suffer from the socioeconomic impacts of gradual shift in the United States to the South and West.

The energy-development process is unlikely to have a neutral effect on employment, income distribution, social and political events, or the environment. A great deal has been said about these phenomena. However, what is still lacking in our national studies is an analytical framework that can be applied consistently to all energy-impacted areas. The ad hoc method of impact assessment prevents the comparison of one community or institution with another, or the tracking of one process through time.

Decision makers in both the public and private sectors must be cognizant of the industrial shift in the United States. The transition from a rural to an urban society, from an agricultural to an industrial economy, and from a blue-collar to a white-collar work force has been tempered by gradualism and close scrutiny. The rapidly increasing concern with energy as a factor of production is a phenomenon of the 1970s, however. Inasmuch as energy influences the amount and location of industry in a given area, it needs to have the same kind of detailed study that other major industrial shifts have received.

MARGINAL ANALYSIS AND LARGE-SCALE ENERGY-DEVELOPMENT PROJECTS

An important group of questions focuses on the decision-making process used by instigators of large capital-intensive projects. Economists have traditionally held that decisions are made at the marginal level. However, with regard to large-scale energy projects, there is some difficulty in determining how long and under what circumstances decisions really remain marginal level and when they become irreversible. Anomalies can arise during a project's development that suggest that marginal decisions play a small or nonexistent role once the project is under way. An example of one of these decision-making anomalies is the continuation of projects that experience labor requirements exceeding 100 percent of the planning estimate.

A reasonable explanation for continuing is that the sheer scale of development imposes an inertia on the project that prevents discontinuance. Many of these projects cost hundreds of millions of dollars, and the sunk costs cannot be easily recouped unless the project materializes. This brings to mind the second plausible explanation.

The price elasticity of demand for most energy sources is very inelastic in the short and intermediate periods. As a consequence, energy producers may be able to adjust the price or rate structure to recoup unforeseen development costs.

Seemingly, once these projects are under way, there is no way to stop them. Even when national priorities change, as they did with regard to the Clinch River Breeder Reactor Project, the project is likely to remain viable while political, technical, and economic forces are mustered in its defense. The energy sector seems to be a fertile ground for research related to marginal decision making in U.S. industry.

THE NATIONAL ENERGY PLAN'S IMPLICATIONS
FOR INCOME DISTRIBUTION AND WINDFALL PROFITS

The president's first guiding principle related to the development of the National Energy Plan was that it be fair. No one can precisely define fairness, but virtually everyone knows when an action or a proposed action is not fair. Nevertheless, every vested interest sees its situation as being unique and therefore exempt from the customary criteria of fairness. The proposition, for example, that "if you want my gas, you're going to pay my price" is suggestive of the attitudes of some groups. Similarly, the proposition that "we want industry, but only if it is clean" suggests that someone else must absorb the pollution. Another attitude of particular pertinence in coal-producing states is the position, throughout the rest of the nation, that "we want your energy, but burn the coal in your state and send us the electricity." These attitudes can only result in the increased balkanization of the nation as we move toward our energy goals.

Income redistribution is at the heart of a major energy controversy over fairness. Capitalizing on the nation's energy problems could hinder income-redistribution goals. By all accepted measures, the distribution of income in the United States is about the same today as it was two or three decades ago. Virtually everyone has more today than he did earlier, and as a consequence, the relative stability in the distribution has been tolerable. The onslaught of the energy problems and their likely impact on economic growth suggest that the economic pie may not be increasing very rapidly. Thus, improvements for one group are very likely to result in negative impacts on others.

If this proposition is basically true, it portends problems ahead. Those at the lower end of the economic spectrum can expect their relative position to deteriorate rapidly as those with control over energy resources improve their position. This is not the forum to examine the marginal propensity of different groups to save, and how this factor affects income changes and employment. There is reason to believe, however, that significant changes in the distribution of income will affect employment. Thus, energy's impact on income distribution could have far-reaching economic consequences for the whole nation.

Windfall profits, by definition, occur without the firm or individual producing more of a product or a better product. In fact, most windfall gains are price phenomena that occur as mismatches in supply and demand occur. Particularly important are rapid price changes that are somewhat predictable. If firms are reasonably sure that energy prices will continue to rise, it makes sense to hold back as much of the resource as possible to capture the higher price later. This activity is a self-fulfilling prophecy of sorts. By holding back on supply, prices may be forced up, and higher profits may be subsequently realized.

It is difficult to determine the relative effects of windfall profits and changing income distribution on employment. Some would argue that higher profits prompt investment expenditures and hence increase employment opportunities. Others may argue that the employment consequences of either phenomenon depend in large part on where the subsequent spending occurs. Some types of investment, that is, petroleum refining, are not very labor intensive, and the job-creating potential may be very small.

In any case, it may be reasonable to suggest that the role of windfall profits in the system may not be uniformly undesirable or, for that matter, avoidable. Walter Mead correctly depicts windfall profits as increases in the value of a firm's inventory, resulting from a fortuitous price increase. He refers to them as a governmental "hangup" and does not feel that the concept is a useful mechanism for determining public policy. His reasons are essentially as follows:

First, for oil discovered after 1974 (when prices reached their initial peak), the term windfall gain would be inappropriate if used in reference to an oil-producing company's inventory valuation. These inventories were not known prior to the price rise and, therefore, could not reasonably be included in windfall profit calculations.

Second, there may be reasonable arguments against inputting windfall profits even for reserves existing prior to 1974. Mead argues that producers may have been expecting oil price increases during the 1950s and 1960s, which never materialized. The abrupt increase in 1973 that resulted in roughly a fourfold increase corresponds to a 3.29 percent compounded annual rate of increase over the two-decade period. This increase is very close to the average real rate of return on capital over long periods of U.S. history. The bunching of the price increases thus creates what he terms a "windfall illusion."

Third, while Mead argues generally for price decontrol, he suggests that if windfall gains are the rationale for control, they should extend to a variety of products other than crude oil. He notes that spot coal prices, Douglas fir timber, and spot uranium (yellow cake) prices have all increased in about the same proportion as oil prices.

Fourth, price controls create distortions, and the longer they are maintained, the greater the distortion, the higher the cost of administration, and the more difficult it becomes to decontrol them. He argues that windfall profits are a signal to the system that these nonrenewable natural resources are becoming increasingly scarce. "Higher prices," he suggests, "are needed to lead people to conserve and to search out substitutes."[10]

It is significant that the president outlined the principles of the NEP to include the following: "The fifth principle is that the United States must solve its energy problems in a manner that is equitable to all regions, sectors, and income groups."[11] The NEP provides further that "the energy industries need adequate incentives to develop new resources and are entitled to sufficient profits to encourage exploration and development of new finds. But they should not be allowed to reap large windfall profits as a result of circumstances not associated with either the market place or their risk-taking."[12]

Equity and fairness are elusive concepts in general, and they are particularly difficult to estimate and evaluate in the complex world of energy, income, and employment. We must be cognizant of these difficulties as policies are developed, but we should not be overwhelmed by them. The NEP will be difficult to implement under the best of circumstances. If Americans feel it is unfair or inequitable, its chances of success rapidly approach zero.

CONCLUSION

In the final analysis, the macroeconomic issues related to energy development and consumption are the most difficult, from an analytical perspective. The reasons are obvious: the macro questions are essentially unconstrained in terms of geographic coverage and time (U.S. energy policy clearly impacts, and is impacted by, energy policy in other countries; many of the macroeconomic impacts on investment patterns and consumption relationships will require decades of adjustment); macroeconomic issues are inextricably intertwined with all social and political decisions, some of which will evolve slowly over the next several decades (income-redistribution policies involving economic and social equity will require many decades of readjustment).

These observations do not suggest that the microeconomic problems are less important, but rather, that the analytical tools are better developed for microanalysis. There are many things that are not understood about microeconomic relationships, but in terms of complexity and temporal difficulties, it seems reasonable to argue that the macroeconomic problems are significantly more difficult.

NOTES

1. Bruce Hannon, "Energy, Labor, and the Conserver Society," Technology Review, March/April 1977, p. 4.

2. Ernst R. Berndt and David O. Wood, "Technology, Prices, and the Derived Demand for Energy," Review of Economics and Statistics 57, no. 3 (August 1975): 260.

3. Ibid.

4. Hannon, op. cit.

5. Ibid.

6. Ibid.

7. Ibid.

8. Ibid.

9. Discussion with Bruce Hannon, Energy Research Group, University of Illinois, November 2, 1977.

10. Walter J. Mead, "An Economic Appraisal of President Carter's Energy Program," Reprint Paper 7 (Los Angeles: International Institute for Economic Research, September 1977), pp. 18-19.

11. Executive Office of the President, Energy Policy and Planning, The National Energy Plan (Washington, D.C.: U.S. Government Printing Office, April 29, 1977), p. 27.

12. Ibid., p. 28.

6
ALTERNATIVE TECHNOLOGIES AND EMPLOYMENT

Prior to late 1973 and, to some extent, afterward, most of what are known as alternative technologies were considered Buck Rogers-type activities by most Americans. Few people understood much about the technology of photovoltaics, or that of biomass conversion, ocean thermal gradients, or geothermal energy. The intense discussion about the alternative technologies had been confined to a few laboratories and a small number of scientists and engineers. Occasionally, technological advances would leak into the newspapers, but for the most part they gained limited attention.

To understand how this new dimension of the energy sector will impact on the structure of employment, a more complete description of alternative technologies must be provided. In the abstract, alternative technology is a very general term applied to any energy technology that is an alternative to what currently exists, that is, oil, coal, gas, nuclear, hydro. In fact, we could argue that esoteric approaches to improving the energy efficiency of the existing technologies could also provide alternatives. A notable example is the magnetohydrodynamic (MHD) program being developed in Montana and Massachusetts.[1] However, for the purpose of this discussion, we will attempt to keep the category of alternative technologies relatively pure by focusing only on those that appear to be technological and resource alternatives to the existing energy sources.

Since virtually all energy comes from the sun, the several technologies encompassing the solar energy system comprise the most important alternative technology.[2] The advent of solar cells of a wide variety of configurations, including the photovoltaic cell, the solar panel (collectors), and systems using devices to trace and concentrate the sun's rays, has captured the interest and imagination of increasing numbers of people. However, in the solar energy sector these prospective systems are simply one part of a larger program. Included in this sector are the biomass-conversion system, wind power, wave power, ocean thermal gradients, and many other systems.

A second alternative to what currently exists is the expansion, on a large commercial scale, of geothermal energy. There are several geothermal systems throughout the world that operate on a commercial scale, but only one exists in the United States—the Geysers, in California. Most experts agree that the geothermal potential is extremely large and could be expanded in some areas of the country. In fact, "The National Petroleum Council has estimated that 19,000 megawatts of geothermal power capacity—all of it in California and Nevada—could be on line in the U.S. by 1985. Other estimates of the potential have ranged up to 132,000 megawatts by 1985, or about 15% of total installed electrical capacity at that time."[3]

Another alternative technology is, of course, nuclear fusion. This approach to energy generation involves the fusing together of light nuclei as opposed to the splitting of nuclei in the fission system. The technology is extremely complex and at present there are no commercial applications possible. However, if the technological advances continue, some scientists feel commercial systems may be possible during the last decade of this century.

Mention was made earlier of the potential role of conservation. Many observers consider energy conservation the single most desirable form of energy "production."[4] Their arguments suggest that it is much less costly to save a quad of energy than it is to produce one. In addition, most of the undesirable side effects of production are eliminated, that is, pollution, resource depletion, transportation problems, and most adverse socioeconomic impacts.

An extension of the conservation idea involves increasing the efficiency of energy use. The Council on Environmental Quality has argued that improving energy productivity will have a beneficial effect on employment and on unemployment. The council reasons that total employment will rise and the unemployment rate will fall "because the energy-producing and energy-intensive sectors of the economy tend to have lower labor intensity."[5] This basic relationship has been studied intensively by the Bureau of Labor Statistics in the U.S. Department of Labor and by the Center for Advanced Computation at the University of Illinois. As noted elsewhere in this study, care must be taken to examine not only the number of jobs but also the type of jobs that are created and destroyed as our energy policy is implemented.

Mention could be made of the potential role of oil shale production, coal conversion, exploitation of geopressured, geothermal zones, and utilization of microwaves to beam energy from solar cells in space, all of which may provide part of our energy needs in the future.

It is important to note, however, that with the exception of the latter system, alternative-technology purists do not see the strategies relying on depletable resources (primarily fossil fuels, but also nuclear fission) as acceptable alternatives in the long run.

One element of the alternative-technology program that is not universal, but is predominant in the overall approach, is the idea of small-scale, decentralized energy systems. The installation of solar collectors on private residences, the utilization of biomass-conversion programs to produce energy and eliminate a city's organic waste simultaneously, and the construction of wind-powered generating stations are all decentralized energy-production schemes. Most of the existing energy sources and several of the alternative technologies (fusion, oil shale, coal conversion) are highly centralized, capital-intensive systems.

The alternative technologies have a wide range of potential employment impacts. Some emanate from the technology itself, some from its deployment (both the time and location), and some from the socioeconomic and legal ramifications surrounding these systems. Unfortunately, the employment consequences of these systems are not well known, and in fact the methodology and data required for their assessment are at a relatively primitive stage of development. * Nevertheless, there are some things that are known, and the pace of development is accelerating. The remainder of this chapter will outline what is known about these relationships. The principal focus will be on the solar energy segment of the industry.

JOBS FROM THE SUN

The implication of this section's title is that jobs flow from the sun. The idea is not as farfetched as it may appear. The substance of the argument is that an aggressive solar energy program is likely to be a net job producer. This is particularly true in the short and intermediate term. A program for producing, installing, and maintaining solar equipment is likely to be more labor intensive on a Btu/labor-unit-input basis than is the displaced current energy source. This presumed favorable relationship is further enhanced by the idea that solar energy is safer, less environmentally damaging; is perpetual

---

*The Bureau of Labor Statistics is attempting to disaggregate the major energy sectors in its macro-forecasting model. This important work will provide an analytical mechanism capable of studying some of the employment consequences of the major alternative technologies.

and decentralized; and permits the shift of fossil resources to better uses. *

How persuasive is the employment argument related to solar energy? The methodology for determining long-run solar energy/employment relationships is even less developed than are the methodologies for analyzing several of the established energy industries. [6] Nevertheless, in the near term some basic relationships have been established, and a review of them will provide some insights into the impact of solar energy development on employment. A basic ratio found in several studies seems to suggest that developing solar hot-water heating and related construction to accommodate solar systems produces three or four times as many jobs as would a hard-technology energy system required to produce the same amount of energy. [7] The basic idea is an intriguing one because of its implicit suggestions related to tax policy, employment and training programs, industrial development and location, and many other factors. It is significant to note, however, that there may not be a one-to-one trade-off in terms of the types of jobs involved; in fact, a critic of this basic idea could reasonably argue that while there are more jobs in total, most of them are lower-skilled and presumably lower-paying jobs. [8] Much would depend on who got the new jobs, who lost the old jobs, where they were located, what types of working conditions accompanied each job type, opportunities for job advancement, job security, and myriad related factors. A cursory review of the relative characteristics of the losers and gainers suggests that solar job creation may compare favorably.

Before looking at several estimates of the employment dimension of the solar industry, a note of caution is in order. It is very clear that solar energy, as we currently think about it, will not displace, on a Btu-for-Btu basis, the conventional energy-producing facilities. Unless dramatic breakthroughs in large-scale electrical storage systems occur, solar systems will require conventional backup systems. We will not see fossil-fueled or nuclear plants disbanded because of the advent of solar facilities. Rather, conventional plants and their distribution networks will remain intact but hopefully will be used less intensively. In this sense, solar development will occur as a net add-on to employment rather than as the substitution of one type of labor for another. Presumably, there will be a slow-

---

*Most energy experts argue that burning fossil fuels for heat, light, or transportation is a relatively poor use of these resources. Presumably, they should be used for chemical feedstocks and in other processes in which there are no viable alternatives.

down in nonsolar development, a gradual phasing out of superfluous plants, and a resultant decrease in the rate of fossil-fuel consumption. These processes will be evolutionary in nature as we develop an increasingly productive solar sector. It is likely that there will be substantial generating capacity (hopefully on a reserve or standby basis) well into the next century.

The estimates of employment impacts of solar development rely on a variety of methodologies and underlying assumptions. Nevertheless, it is significant that they all come out in about the same place. Representative of these studies is the one recently prepared by the California Public Policy Center (CPPC), entitled Jobs from the Sun. [9] This effort focused on the employment consequences of solar energy for space and water heating. Clearly, if conditions and technology warrant rapid expansion that includes photovoltaics, wind, bioconversion, and other systems, the job effect will be much greater.

Over the ten-year interval (1981-90), if Californians used existing solar technologies, they could create over 375,000 jobs per year in this industry. [10] Since the CPPC analysis focused on the "minimum development scenario," this estimate of employment may be on the low end of the continuum.

Using solar energy to replace or supplement fossil fuels for space and water heating will produce several other benefits as well. They include:

A $42 billion increase in personal income over the decade,
A $51 billion increase in gross state product,
A $19.8 billion tax savings, and
A $10.2 billion savings of exported capital. [11]

It is difficult to extrapolate the California estimates to generate national estimates, because of the wide variability of climatic conditions, legislative systems, economic conditions, availability of materials, and public attitudes. Nevertheless, it is interesting to note that about 10 percent of the nation's population resides in California. Inflating the employment, personal income, tax savings, and related data by a factor of ten results in a dramatic overview of solar energy's potential: 3 or 4 million new jobs each year could virtually eliminate our current unemployment problem. In addition, these jobs would presumably be geographically dispersed. Therefore, they would minimize the bottlenecks encountered in highly centralized development projects.

The potential employment in a developing solar industry is substantial. If a factor of only five were applied to the California experience, the impact on the economy would be dramatic. It is important to reiterate that the California study presumably estimated the low end of the potential solar development. It may be reasonable to suggest that a much larger potential impact exists.

A second analysis was prepared by the Office of Technology Assessment (OTA) for the administration's Domestic Policy Review of Solar Energy.[12] The OTA compared a hard technology system comprised of an 800-megawatt, coal-fired electricity-generating plant and two types of solar systems, that is, solar hot-water heaters and a tracking silicon photovoltaic system. The comparisons were in terms of man-hours per megawatt-year to ensure that the labor requirements were on an equivalent basis. For reasons not specified, the coal-fired plant had a 30-year life while the solar systems were assumed to operate for only 20 years. This difference prevents a strict year-to-year comparison.

In any case, the labor requirements for these systems on a megawatt-year basis are instructive. The coal-fired plant was estimated to require 2,348 man-hours per megawatt-year for the construction, operation, and maintenance of the plant over its expected 30-year life. The hot-water solar system was estimated to require between 3,540 and 5,240 man-hours per megawatt-year for all three functions. The photovoltaic tracking system required between 11,200 and 15,040 man-hours per megawatt-year, depending on the assumptions made about a backup capability.[13] Clearly, the solar options are significantly more labor intensive than is the coal system.

It is more difficult to extrapolate these types of estimates to national totals because of the required assumptions about the speed of development, the particular solar technology mix utilized, and the public acceptability of the solar option. What is significant, however, is the basic set of ratios apparent in this analysis. While not exactly parallel to the California estimates, they are of the same magnitude.

A third study of particular relevance to the subject matter of this chapter was outlined in the testimony of James Benson, before the Energy Subcommittee of the Joint Economic Committee, concerning the employment consequences of alternative energy futures.[14] Benson attempted to demonstrate that the employment-generating potential of a large conservation/solar package was several times as large as that of a nuclear facility on an equivalent Btu basis. Specifically, for the programs specified in the analysis, the following relationships occurred:

Phase	Conservation/Solar Package (man-years)	Nuclear Project (man-years)
On-site construction	75,120	27,880
Direct manufacturing	36,350	11,030
Operation/maintenance	66,880	27,700
Total	178,350	66,610

From these estimates, Benson concluded that the conservation/ solar approach produces about 2.7 times more employment than the nuclear option. [15]

Another way of looking at the relative labor intensity of the two options is to examine the man-years of work per million dollars of production cost. The conservation/solar package is estimated to produce 30.2 man-years of work per million dollars, while the nuclear option produces 9.8 man-years of work. On an equivalent dollar basis the conservation/solar option is even more labor intensive. *

During the same hearings, Bruce Hannon of the Center for Advanced Computation at the University of Illinois, provided evidence fully supportive of Benson's principal findings, but suggestive of a somewhat different energy strategy. [16] Hannon argued that major employment impacts can be produced by shifting the structure of various economic activities. For example, he argued that changing from intercity-plane to intercity-train travel would result in 930,000 new jobs for each quadrillion Btu's of energy saved. Implementing this change in the transportation system, which would reduce overall energy consumption by approximately 5 percent, would presumably move the economy to full employment.

Other changes are less efficient in terms of creating new jobs in relation to energy conservation, but several are nonetheless potentially important sources. For example, changing from intercity-car to intercity-bus transportation creates about 330,000 new jobs per quad of energy saved; changing from frost-free to conventional refrigerators produces about 60,000 jobs per quad saved; and changing from electric to gas-water heaters would produce about 120,000 new jobs per quad saved. While no one of these changes can be relied on to solve the unemployment problem, clearly, policy makers should be cognizant of the differential employment impact of their actions. If options are essentially balanced in their other advantages and disadvantages, the job-creation potential of the effort should be weighed heavily. A comprehensive strategy would probably encompass many of the changes studied by Hannon.

SOLAR RETROFIT VERSUS INSULATION

The discussion about the use of alternative technologies ultimately focuses on what the optimum strategy should be for imple-

---

*As we noted earlier, the types of jobs may be just as important as the total number of jobs. Much more work is needed to determine the skills mix.

mentation in the short and the long run. Some observers believe that the United States should move ahead rapidly to retrofit homes, factories, and commercial buildings. Others want to retrofit where possible, but to ensure that new structures use passive and active solar systems to the maximum extent possible. The latter position is becoming the increasingly popular one because of the potential costs and benefits of solar retrofitting as opposed to simple insulation. Preliminary work by Frank Hopkins of the Department of Energy shows that the potential benefits from solar retrofit in most structures may not justify the costs. In most instances, it would be more cost effective simply to insulate the structure to a high level of heat resistance (which must be done prior to solar retrofit, too) and reap the energy savings that occur. To go to the next step and retrofit the structure, that is, produce some energy, may not justify the additional cost. As a consequence, the implications of this analysis are that, in terms of a short-term national strategy, it may make more sense to develop a large-scale insulation program for existing structures and retrofit only where there are clear-cut opportunities for positive benefit flows. Every effort should be made, of course, to ensure that solar applications are utilized in new construction.

Proponents of solar technology may interpret the results of this analysis as antisolar. Nothing could be further from the truth. The fact of the matter is that solar-retrofit technology is relatively poorly developed and is relatively expensive. It is not at this point a very efficient energy producer. This situation will change as new technological breakthroughs occur. The analysis supporting increased insulation as opposed to solar retrofit simply acknowledges the fact that we are more proficient at saving energy through conservation than at producing it through solar retrofit. Therefore, on a resource-allocation basis, it may make more sense to channel resources into insulation programs rather than solar-retrofit programs.

It is very important to note also that this policy suggestion does not argue for abandonment of the solar-retrofit idea. Resources should be allocated to research and development (R&D) efforts to improve the cost effectiveness of solar retrofit. At some point there may be a reversal of the relative effectiveness of the two approaches.

From the standpoint of their employment impacts, both insulation and retrofit programs are relatively labor intensive. Certainly, a retrofit program would provide more employment opportunities for sheet-metal workers and plumbers, while an insulation program would assist carpenters and laborers. On the production end, different industries and occupations would be impacted. Overall, the net employment differential between the two strategies is likely to be small. As a consequence, the selection of the strategy mix would probably not revolve around the relative job-creating capacity of the two alternatives.

SUMMARY

The exploration of alternative energy sources involves a vast array of complex technological, resource, economic, and social factors. No one disputes the fact that we can generate useful energy from the sun through simple solar energy collectors, photovoltaics, wind, bioconversion, ocean thermal gradients, and several other technologies. A small, but growing, amount of energy is currently flowing from these sources. The debate focuses on how to make these various sources efficient enough so that they become viable competitors or replacements for the traditional energy sources. Within the parameters of the debate are a multitude of competing concerns: the environmental impact of the options; how they will affect capital and financial markets; how they will interface with the existing systems; whether they are socially acceptable; whether the technology is sufficiently developed to warrant commercialization; and how each option will affect employment and the distribution of income.

Every one of these issues is a legitimate area of concern and discussion. It may be impossible to reconcile fully every issue, but the potential process must ultimately derive a consensus that is in the public interest.

No one seriously argues that the employment dimension of the debate is considered the single most important issue. Indeed it may or may not be. What is important is that the employment issue receives its deserved attention. There is growing evidence that the job-creating ability of solar and conservation programs is much greater than that of the alternative options. No one seriously doubted that relationship, but empirical evidence is bolstering intuitive judgments.

Senator Edward M. Kennedy has argued that the Congress should not debate energy policies without having access to this important body of information. He was not suggesting that every decision would rest on its relative employment consequences, but that these were important considerations in the political calculus. The analysis thus far has been useful in terms of providing valuable insights and general guidelines for discussion purposes. What is needed now are more in-depth analyses that examine the likely impacts along occupational, industrial, geographic, and temporal lines. These analyses must be based on methodologies that are capable of capturing the dynamics of a rapidly changing national and international economy.

The next chapter will outline three current modeling programs that are seeking better methods of analyzing various segments of the energy industry. What will become apparent is that these modeling efforts, while important parts of the overall energy-sector analytical program, provide few insights into the character of the employment changes in the alternative-technology sectors. These modeling activities have made large gains in the last two or three years, and at some point there may be an integration of some or all alternative technologies into the analytical framework.

NOTES

1. The MHD technology may yield up to 50 percent more usable energy from each unit of coal burned, when compared with the conventional coal-fired generation plant. See Energy Research and Development Administration, A National Plan for Energy Research, Development and Demonstration: Creating Energy Choices for the Future, vol. 2, "Program Implementation," ERDA 76-1 (Washington, D.C.: U.S. Government Printing Office, 1976), pp. 33-34.

2. Henry Simmons noted that "in theory, an earth-based solar collector 1/500th of the area of the United States (an area slightly smaller than Massachusetts) receives an amount of solar energy that if converted at 20% efficiency, would provide for all of the Nation's present consumption of electricity." Henry Simmons, The Economics of America's Energy Future (Washington, D.C.: Energy Research and Development Administration/U.S. Government Printing Office, 1975), p. 45.

3. Ibid., p. 43.

4. Many critics are quick to point out that while conservation is important in principle, how it occurs, that is, through higher prices or technological change, makes a difference, as does who absorbs its impact, that is, the poor and elderly, everyone, the industrial community, or any other identifiable group. The Oil, Chemical, and Atomic Workers International Union (OCAW) maintains the position that "higher prices and the hardship they imply are not the answer to present problems of energy misuse and waste." The OCAW prefers tax breaks or low-cost loans to encourage conservation efforts. See Oil, Chemical, and Atomic Workers International Union, OCAW Energy Policy (Denver: OCAW International Executive Board, May 1977).

5. Council on Environmental Quality, The Good News about Energy (Washington, D.C.: U.S. Government Printing Office, 1979), p. 34.

6. The Office of Technology Assessment recently noted that "long-term labor impacts will depend on forecasts of growth rates both in the economy and in U.S. energy consumption—subjects about which there is great confusion and disagreement." Office of Technology Assessment, Application of Solar Technology to Today's Energy Needs, vol. 1 (Washington, D.C.: U.S. Government Printing Office, June 1978), p. 210.

7. The Employment Impacts Task Force of the Impacts Panel, of the Domestic Policy Review of Solar Energy, concluded in a draft report, "A Response Memorandum to the President of the United States" (Washington, D.C., September 21, 1978), that "solar heating systems provide about 3 to 4 times as many direct manufacturing jobs as conventional electric power for an equivalent amount of energy production" (p. 42).

8. The California Public Policy Center noted that solar jobs "demand a relatively low level of skill, and would thus be particularly well suited for areas of high unemployment." California Public Policy Center, Jobs from the Sun: Employment Development in the California Solar Energy Industry (Los Angeles: CPPC, February 1978), p. 10.

9. Ibid.

10. Ibid.

11. Ibid., p. 1.

12. Office of Technology Assessment, op. cit.

13. See ibid., particularly pp. 210-13.

14. James W. Benson, Energy and Employment testimony before the Energy Subcommittee of the Joint Economic Committee, 95th Cong., 1st sess., 1978.

15. Duane Chapman suggested supporting analyses in "Taxation, Energy Use, and Employment," testimony before the Energy Subcommittee of the Joint Economic Committee, 95th Cong., 1st sess., 1978. See, particularly, the discussion on pp. 4-5 and table 3, in which the employment consequences of solar and electric water- and space-heating systems are compared.

16. Bruce Hannon, "Conserving Energy While Also Increasing Employment," testimony before the Energy Subcommittee of the Joint Economic Committee, 95th Cong., 1st sess., 1978.

7
DEVELOPING ENERGY/EMPLOYMENT MODELING CAPABILITY

The oil embargo of 1973 touched off concern about the labor market dimensions of shifts in energy production and consumption. Since that time, analysts have studied the employment requirements related to the president's NEP, the short-run disruptions caused by natural gas shortages, and the skills-training requirements related to emerging technologies such as solar, geothermal, and nuclear fusion. All of these efforts to understand energy/employment relationships are of value in providing a frame of reference or a perspective on the magnitude of the problems.

However, virtually all of these studies on energy and employment have had serious limitations. The most important of these limitations is that they provide estimates that are too aggregated to be useful for most planning purposes. It does a coal operator in Price, Utah, or in Gillette, Wyoming, little good to know that at the national level there is no shortage of coal miners if he cannot hire miners locally. Nearly everyone agrees that, in the aggregate, a shortage of workers in the energy industries is unlikely. However, there will very likely be spot problems involving workers in general or in certain occupational areas. The energy/employment modeling program must be capable of identifying these supply and demand imbalances.

There has been a tendency to focus on direct employment in energy industries. However, there are indirect effects, price and income effects, and structural changes involving the occupational mix that must be analyzed. Admittedly, examination of these additional dimensions of energy/employment relationships is fraught with theoretical and quantitative difficulties. Nevertheless, it can be argued that the indirect effects and the price and income relationships will have the largest impact on employment in the long run. A comprehensive, coherent energy/employment policy cannot be structured without the inclusion of these relationships.

The purpose of this chapter is to review several energy/employment models that are currently being developed.[1] It would be very difficult to analyze all models that address these relationships directly or indirectly. In fact, virtually every large macro model, for example, the Data Resources Incorporated (DRI), Wharton, Chase Econometrics, has an energy subsystem that can generate aggregate energy/employment data. These data are useful as general indicators of how the economy is structured and what basic changes are occurring, and can be used as control totals for subnational models.

Three models were selected as reasonably representative examples of what is being done and where new initiatives are likely to occur. The first model, the Construction Labor Demand System (CLDS), is being developed under an interagency agreement between the Department of Labor, the Tennessee Valley Authority, and the Department of Energy. The second model, the Regional Industrial-Occupational Labor Demand (RIOLD) model, is being developed by the Bureau of Economic and Business Research at the University of Utah, for the region 8 Employment and Training Administration Office of the U.S. Department of Labor. The third model, the Energy Supply Planning model (the Bechtel model), is a Bechtel Corporation project under the sponsorship of the National Science Foundation.

The characteristics, objectives, and stage of development vary significantly among these three systems. Each model adds to the state of the art, and there is relatively little duplication of effort (there is some overlap in the CLDS and Bechtel models). While we believe these three modeling programs are useful, we are not endorsing them as necessarily the best ones available or suggesting that other initiatives are unnecessary. The three systems are in various stages of development, and at this juncture they appear promising. As experience is gained with these systems, we will be in a better position to evaluate their effectiveness.

CONSTRUCTION LABOR DEMAND SYSTEM

During 1976/77 four federal agencies (the Department of Labor, the Federal Energy Administration, the Energy Research and Development Administration [ERDA], and the Tennessee Valley Authority) cooperated in the design and development of the CLDS, a system for estimating the manpower requirements in the energy sector, by occupation, for known construction activities. These estimates will be aggregated at the county, the Bureau of Economic Analysis (BEA), the standard metropolitan statistical area (SMSA), the state, and the national levels.[2] The output from this system will be compatible with the other components of the CLDS program (commercial and industrial building, nonbuilding construction, residential construction, and institutional construction) in terms of occupational data (29 craft categories), geographic location, time horizon (monthly for a two-year horizon), and by activity (constant dollar value of construction units).[3]

Description of the Model

The CLDS energy-sector model tracks and estimates employment requirements in seven categories:

1. electricity-generating facilities,
2. transmission lines,
3. coal,
4. uranium,
5. oil,
6. gas, and
7. other energy sectors (solar, geothermal, R&D projects).[4]

The system produces time-phased output showing the requirements for 29 occupational classifications during the entire construction phase. The occupational composition of the work force varies over time. For example, "A larger proportion of iron worker manpower requirements are reenforcing iron workers in the early project stages and structural iron workers in the later stages."[5]

The CLDS energy-sector model is being developed in two phases. "Phase I will implement the basic methodology on a simplified scale, incorporating all of the basic forecasting, coverage, conversion and statistical analysis techniques to be used on the more sophisticated Phase II effort."[6] The Phase II effort "will expand data coverage, incorporate higher levels of detail, and include the integration of econometric estimating techniques."[7]

The model will generate short-term forecasts (zero to three years), mid-term forecasts (one to ten years), and long-range forecasts (zero to 25 years). Obviously, as the time horizon is extended, the confidence of the estimates decreases.

The CLDS will utilize a variety of data sources for building up the data base and for periodic updating. Since there is no single source of data for all types of construction, integration of several rather diverse sources is required. The major sources are:

1. the Federal Power Commission's Generating Unit Reference File (for coal-, oil-, and gas-generating facilities),
2. ERDA's Civilian Enriching Requirements Program (for nuclear facilities),

3. direct surveys of hydroelectric plants,
4. Contractors Mutual Association,
5. F.W. Dodge reports,
6. FEA, Utility Project Operations Monthly Report 1960-1977,
7. National Technical Information Service, Construction Status Report: Nuclear Power Plants, July 1977, and
8. Department of Commerce, Construction Review: 1960-1970.

CLDS staff members believe that the data base from which the manpower estimates are derived is the most comprehensive assemblage of energy/employment data anywhere in the nation.

The conceptual process underlying CLDS is relatively simple. The first step is to develop "demand forecasts (in dollars) for the various construction sectors."[8] The second step involves "conversion of dollar demand by construction sectors into labor requirements by craft."[9] The system estimates the monthly flow of various manpower types as the construction projects progress toward completion. In other words, a time profile of manpower demand by craft is the immediate product of the estimation process. An important element in this segment is the input from the BLS construction surveys. "Since 1958, BLS has been conducting regular surveys of man-hours by trade per constant $1,000 for selected types of construction."[10]

It is significant that the Phase I estimates are based on a fixed-coefficient model. Since Phase I only projects in the short or intermediate terms, the assumption that technology and factor prices will be constant is a reasonable compromise made to simplify the model. The Phase II segment will contain models that deviate from the fixed-coefficient assumptions. The Phase II model must reflect "(a) changes in the relative prices of labor, materials, energy and capital; (b) changes in safety and health requirements as expressed in building codes; and (c) changes in the scale of projects."[11]

The CLDS is, as the name implies, a labor-demand estimating mechanism. The developers of CLDS have initiated a labor-supply project, but it may take several years before this project is completed. Recognizing the difficulties of working on the supply side, CLDS designers believe that "it will be impossible to provide labor supply data of such quality and quantity in order to give users a balance sheet approach to evaluating labor demand and labor supply in a given area."[12] Several agencies and institutions, including the Nuclear Regulatory Commission, the Tennessee Valley Authority, and the University of Utah, are working on labor-supply projects that may ultimately have important implications for the CLDS project.

THE REGIONAL INDUSTRIAL-
OCCUPATIONAL LABOR DEMAND MODEL

The RIOLD modeling program was initiated in 1976 because of the dissatisfaction with existing regional modeling efforts and particularly their application to the energy sector. A variety of regional prototype models were evaluated during the early design and development stages, but none appeared to provide sufficient structural adaptability to satisfy the needs of the regional planners.

The RIOLD model is an econometric modeling system that "incorporates a dynamic industrially disaggregated formulation of the traditional economic base concept."[13] While the impetus for the model's development process was the emerging energy sector in the Rocky Mountain region, it is important to note that the RIOLD has general applicability. The model can be used "to produce industry and occupation employment impacts for any change in basic sector activity."[14]

The model is structured to provide labor-demand estimates in geographic subdivisions called labor market areas (LMAs), as defined by the Department of Labor. The general description of an LMA is a geographic area comprised of one or more counties in which there is a concentration of economic activity or labor demand, such that workers can readily change jobs without changing their place of residence. The RIOLD is capable of estimating the timing as well as the magnitude of employment changes, so that a time path of development can be determined.

The model has the capability to produce current and projected employment estimates in 76 industries and 420 occupations. The 41 basic industrial sectors contain 10 directly involved in energy production, processing, or transportation. The 10 energy sectors are coal mining, oil shale mining and extraction, petroleum and gas extraction, pipelines, petroleum refining, electrical power generation, coal gasification, power plant construction, coal-gasification plant construction, and oil shale plant construction.

Clearly, there is some spillover into several other sectors, such as electrical machinery and railroad transportation, but for purposes of analysis, the ten basic energy sectors will contain virtually all of the employment changes.

Description of the Model

This combination econometric/economic base model is calibrated with "county specific historical data on wages and salaries, earnings and employment from the Regional Economic information

System (REIS) file of the U.S. Department of Commerce, Bureau of Economic Analysis (BEA), and region specific industry-occupation matrices developed within the framework of the National/State Industry-Occupation Matrix System, conceived by the U.S. Department of Labor, Bureau of Labor Statistics (BLS)."[15] The two-digit SIC scheme in the BEA income and employment data determined the level of detail within the RIOLD model. Data from the BLS Industry-Occupation Matrix System were added to the BEA categories, thus giving 70 industrial sectors. As noted earlier, several of the energy sectors were further disaggregated to provide a total of 76 industrial sectors. The 420 occupational categories come directly from the BLS Industry-Occupation Matrix System.

The 76 industrial classifications have been grouped, for analytical purposes, into 35 residentiary sectors (producing goods and services principally for consumption by area residents); 33 basic sectors (primarily export dependent); and 8 sectors that are some combination of basic and residentiary.

The architects of the model indicated that "the method employed in the construction of the model involves an industrially disaggregated formulation of the conventional basic-residentiary multiplier approach in conjunction with an occupational requirements sub-model."[16] This model is different from other regional modeling efforts in that the parameters are statistically estimated on the basis of empirical information in the BEA's REIS file and the BLS regional industry-occupation matrices.

The model raises the timing of development to a position of central importance by allowing the distributed-lag characteristics of the model to infer the time path of development as well as the full-equilibrium level of development.[17] The parameter estimates are based on both cross-sectional and time-series aspects of the BEA earnings and employment data. Several energy and construction sectors have time-phased occupational-requirements vectors as compared with the static vectors in the regional industry-occupation matrices.

The basic inputs to RIOLD are the projected employment changes in the primary (basic) energy-producing industries.[18] It is important to reiterate that the model will accept projected employment changes in other basic industries as inputs and thus "produce occupational projections for all types of industrial expansion."[19]

A wide variety of data sources have been developed to determine energy-development programs by type of industry, location, size, length of construction period, and, if available, employment requirements for each part of the construction and operation phases. To determine the changes in employment, by year, in the energy sectors, the data were compared with the 1977 actual employment benchmark data.

The First Run of the RIOLD Model

The first run of the RIOLD model occurred in April 1978. One LMA in each of the region-8 states (Colorado, Utah, Wyoming, North Dakota, South Dakota, and Montana) was selected for study. This program was designed to demonstrate the model's capability and provide illustrations for interested groups.

The first complete run occurred in August 1978, covering all energy-affected non-SMSAs in region 8. Included in the run were 19 areas encompassing 97 counties. The data are still being analyzed and most of the data are not yet available for wide distribution. However, several interesting, generalized observations can be made. For example, coal mining jobs in the region will more than double between the base year (1977) and 1983. The model estimates an increase of about 21,900 jobs between 1977 and 1983. About 1,900 new workers will be required to operate new coal-fired power plants.

The model produced employment multipliers (the ratio of total jobs to energy jobs) that average 2.13 between 1977 and 1983. This means that for every job created in the energy sector, an additional 2.13 payroll jobs will be created in the secondary sector. As expected, the multipliers were significantly lower in transitory basic-growth situations, that is, construction plant operations.

The planned and proposed regional energy development will produce about 62,000 new jobs between 1977 and 1983. This estimate includes both the primary and secondary changes.

The RIOLD model is an innovative effort to provide dynamic estimates of employment changes at a high level of geographic disaggregation. While there are a number of regional models in existence, most experts agree that few adequately reflect the employment dynamics of small economic areas. It is too early to judge the RIOLD model a success, but the validation process seems to suggest that it is working reasonably well. The model's architects are engaged in several refinements, including complete inclusion of the dynamic lag formulation, calibration for metropolitan LMAs, and incorporation of occupational detail from energy-specific projects from the Occupational Employment Survey (OES) program,[20] which should enhance the model's accuracy and timeliness.

THE ENERGY SUPPLY PLANNING MODEL
(BECHTEL MODEL)

The Energy Supply Planning model was developed by the Bechtel Corporation under the sponsorship of the National Science Foundation. It is designed to evaluate "the direct resource requirements for capital, manpower, and materials of alternate energy system development

programs. "[21] Based on a user-specified fuel mix and energy-demand
projections through 1995, the model specifies the number and location
of new energy and transportation facilities needed to accommodate
these energy flows. Schedules of annual resource requirements can
be compared with supply estimates to determine those impediments
that stand in the way of specific energy goals.

Description of the Model

    The model can be disaggregated to 14 geographic regions (North-
east, Mid-Atlantic, South Atlantic, East North Central, East South
Central, West North Central, West South Central, Northern Mountain,
Southern Mountain, Northern Pacific, Southern Pacific [including
Hawaii], Gulf Coast, Atlantic Coast, Pacific Coast). The regional
estimates will permit a more intensive examination of the transpor-

FIGURE 4

Schematic of National Energy Supply Planning Model
(Bechtel model)

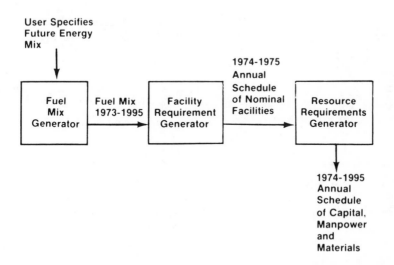

    Source: M. Carasso, "The Bechtel/NSF Energy Facilities and
Resources' Model," mimeographed (San Francisco: Bechtel Corporation, Energy Systems Group, January 28, 1975).

FIGURE 5

Fuel-Mix Generator
(total electric power requirements)

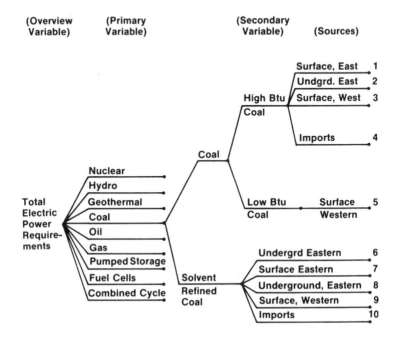

Source: M. Carasso, "The Bechtel/NSF Energy Facilities and Resources' Model," mimeographed (San Francisco: Bechtel Corporation, Energy Systems Group, January 28, 1975).

tation requirements associated with a specified national energy program and a variety of interregional fuel-flow patterns.

The model examines the direct requirements for various energy options. Conceptually, it would be able to identify potential production bottlenecks, examine gross energy efficiency in the system, determine the indirect requirements of a given energy strategy, and facilitate the feasibility of alternative energy programs.

Figure 4 shows a simple schematic of the model. The model's "Fuel-Mix Generator" specifies all energy flows between 1977 and 1995 in a 248×23 matrix. An example for electric power requirements is given in Figure 5.

The "Facility Requirements Generator" estimates the additional facilities required to achieve the target energy flows. This segment of the model examines the capacity of existing facilities, the number and timing of facility retirements, and planned additions.

Based on the schedule of new additions determined in the Facility Requirements Generator, the "Resource Requirements Generator" determines the time-phased flow of manpower, capital, and materials for specific facilities or the entire energy-supply system.

CONCLUSION

The preceding discussion attempted to explain how the three models are structured and, in general, how they function. It would be inappropriate to suggest how useful or effective they will be, because considerable development work remains. However, it may be useful to examine briefly what criteria can be employed to evaluate these and other systems.

The first and, in our judgment, the most important feature of an energy/employment model is its ability to produce disaggregated data. Data must be disaggregated in at least three dimensions: occupation, industry, location. We cannot overemphasize the importance of this criterion, from the perspective of policy development and labor market intervention strategies.

Second, the output must be timely. The energy-development process is dynamic and susceptible to rapid changes in direction. A system that contains a six- or nine-month lag between data collection and report generation is likely to find itself continually too late in providing meaningful input.

Third, the model must be able to provide data that are easily understood by all users. A policy maker who must sift through reams of printouts or understand the differential calculus before he can discern the significance of the report is unlikely to utilize the system. We recognize that energy problems are complex and that simplicity should not take precedence over accuracy or relevance. Nevertheless, the best, most sophisticated model will have minimal impact on policy development if those in policy positions cannot readily understand the message it conveys.

The fourth criterion suggests that the model must be easy to update. Obviously, to have timely information, efforts must be made to update the data base as soon as new data become available. Very complex models are expensive to operate and maintain. As a consequence, there is some reluctance to update them until a basic change in the economy is apparent. Different models serve different purposes and, as a result, have different updating requirements. But our observations suggest that smooth updating generally produces timely output and therefore increases the relevance of the data.

Other factors, such as cost, method of data processing and report generation, and sources of data inputs are also important. At this point, however, it would be premature to evaluate the three systems in terms of these criteria. After the systems have had reasonable operating experience, an intensive evaluation would be appropriate.

The importance of constructing efficient models cannot be overstressed. The next chapter carries the discussion one step further by outlining the research needs in the energy/employment areas. While additional model building or improvement will be required to address several of the suggested research topics, we are convinced that current modeling programs can go a long way toward producing insights into most of them.

NOTES

1. For those who are interested in a review and comparison of other modeling efforts, two recent papers provide an excellent overview. See Albert J. Eckstein and Dale M. Heien, "A Review of Energy Models with Particular Reference to Employment and Manpower Analysis" (Report prepared for the Employment and Training Administration, U.S. Department of Labor, Washington, D.C., March 1978); and CONSAD Research Corp., "Review of Employment/Energy Economic Analysis Methods" (Report prepared for the U.S. Department of Labor and the U.S. Department of Energy, Washington, D.C., January 1979).

2. PRC Data Services Co., "CMDS Conceptual Design" (Report prepared for the U.S. Department of Labor, Washington, D.C., December 10, 1976), p. I-7.

3. Federal Interagency Construction Task Force, Energy Sector, "Construction Manpower Demand System: Energy Sector" (Washington, D.C.: Energy Sector Office, January 1977), p. 1.

4. Ibid., pp. 1-2.

5. Ibid., p. 13.

6. PRC Data Services Co., op. cit., p. I-3.

7. Ibid.

8. U.S., Department of Labor, "CMDS Workshops," mimeographed (Knoxville, Tenn.: DOL, November 4, 1977), p. 12.

9. Ibid.

10. Ibid., p. 21.

11. Ibid., p. 26.

12. Ibid., p. 56.

13. U.S., Department of Labor, Employment and Training Administration, region 8, "Regional Industrial-Occupational Labor Demand Model Description," mimeographed (Denver: DOL, n.d.), p. 1.

14. Ibid., p. 11.

15. Ibid., p. 1.

16. Ibid., p. 4.

17. Ibid.

18. Ibid., p. 5.

19. Ibid.

20. Ibid., p. 11.

21. U.S., Department of Labor, "CMDS Workshops," p. 78.

8
THE RESEARCH AGENDA

This chapter constitutes our attempt to identify those energy/ employment areas in which information is lacking. We do not wish to suggest that the research topics we propose are new, or that nothing is known about them. On the contrary, subjects such as training, mobility, price changes, and long-range employment estimation have been researched in depth. Nevertheless, the energy/employment relationship is a particularly critical interface that requires more consideration than has been given to it in the past.

It is important to point out that several energy/employment problems are regional in character. For example, the problem of the boomtown environment in terms of labor utilization is probably unique to the Rocky Mountain region and the Gulf Coast states. There may be some evidence of boomtown growth in Appalachia, but the predominant problems are centered in the West and the Gulf Coast. However, despite their regional character, studies of boomtown developments are important on a nationwide basis because of the implications they hold for major development projects.

The order of presentation here does not suggest the priority of each topic. Some are obviously more pressing than others, and so we have given them a research-priority ranking. The ranking is on a scale of 1 to 10, with number 10 representing the most important topic. Furthermore, we do not suggest that a single organization or individual attempt to address a specific agenda item. Since many of the problems are complex and require a substantial resource commitment, it is hoped that a coordinated research effort by several organizations will emerge. No attempt will be made to propose how much money is needed to research a specific subject. Resource commitment is contingent upon the length of time committed to the project, staff requirements, geographic coverage, and many other factors.

THE AGENDA ITEMS AND
THEIR RESEARCH-PRIORITY RANKINGS

The Impact of Population Changes on Productivity, Turnover, and Absenteeism [7]

Rapidly expanding energy labor markets generally depend on in-migration as a major supply component. The phasing of a development effort from initial exploration, to construction, and to operation and maintenance results in a rapid influx of new temporary construction workers, a subsequent decline in the construction work force, and a long-run stabilization of the permanent operation/maintenance (O/M) work force. The typical pattern for most projects is depicted in Figure 6. The phase Ot_1 represents new employees involved with initial exploration and site evaluation. Once construction is started, the work force rapidly expands to some maximum level, at t_2 in this case. As construction nears completion, operation and maintenance workers

FIGURE 6

Profile of Construction Workers' Utilization

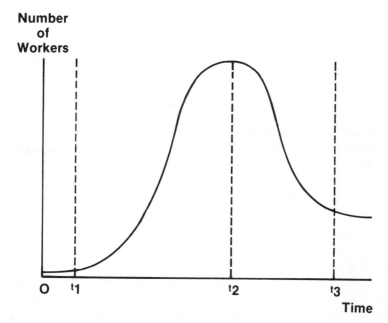

Source: Derived from Joe Garrett Baker, Labor Allocation in Western Energy Development, Monograph no. 5 (Salt Lake City: University of Utah, Human Resource Institute, 1977).

begin arriving. At t_3, most construction workers have left, and the O/M work force stabilizes and continues at a somewhat constant level into the future.

The pattern of expansion creates major problems for a community (particularly a small community) if the project is very large. Area planners can be fairly confident about the work force after t_3, in terms of the required social infrastructure (roads, water, schools, hospitals, parks, and houses). However, accommodating the t_2-level work force presents a major problem. Some construction workers bring their families with them, while others do not. The cultural and recreational requirements of the construction work force and of the indigenous population frequently do not coincide. Providing adequate housing, schools, hospitals, water, sewers, and parks for a temporary work force (which may mean three to five years) is a severe planning problem.

To the extent communities are able and willing to expand their infrastructure, the adverse work force-related problems can be mitigated. When infrastructure development falls short of total needs, the work force will be negatively affected. Contemporary workers are deeply concerned about the adequacy of social overhead capital. This concern affects their performance on the job, determines how long they will stay in a community, and influences absenteeism rates. In addition, the impacts of rapid energy development on the indigenous work force involved in nonenergy activities must be studied.

Some socioeconomic research has focused on these types of problems, but a quantification of the steps that lead to effective remediation has not yet been done. We do know something about how living conditions affect worker productivity. However, the importance of thoroughly understanding the interactions among such variables as absenteeism, productivity, and turnover, as well as work force levels, population levels, and infrastructure requirements cannot be overstressed.

Occupational-, Geographic-, and Industrial-Mobility Patterns in Energy Industries [7]

The movement of workers into an area in response to differential economic opportunities, the movement of workers between energy and nonenergy firms, and the change in occupational status due to training, education, or innate ability are phenomena that are not well understood. The design of employment strategies at the state and substate levels requires very precise information on worker flows. Since all mobility dimensions impact labor-supply estimation, it is extremely important to understand them fully and to have an estimating method. The mobility issue is so critical, from the standpoint of all socioeconomic planning, that a comprehensive analysis cannot proceed very far until these relations are clearly understood and quantified.

Measuring the Flows of Workers into and within the Local Labor Market [7]

Relatively little is known about the timing of the flow of workers into an expanding labor area or, for that matter, out of a contracting one. What determines when workers move? How many respond to direct job offers as opposed to the magnetism of the boomtown phenomenon? What incentives draw workers to the secondary and tertiary sectors? What can community leaders do to increase, decrease, or rationalize the flow of workers into their area? Are there major seasonal or geographic factors that determine mobility? We need a series of intensive case studies focusing on these questions.

The Carryover of Energy Facility Workers into Operational and Maintenance Occupations [5]

In some energy industries, such as coal mining, the workers who develop the mine are frequently the same ones who end up operating and maintaining it. In other industries, however, builders and operators are rarely the same workers. The construction work force that erects a power plant usually does not participate to a large extent in the O/M phase. However, there are many variables that influence the transition process and determine whether or not a worker will stay in an area. Training, wages, general economic conditions, age, and family size all impact the transition process. Information on the carryover of construction workers into operational and maintenance positions can be relatively important in some situations. In general, however, this information is reserved for the nice-to-know category. Certainly, transition patterns can influence labor supply, but the estimation method for labor supply may be able to pick them up.

Examination of Problems of Transition into Energy Occupations of Nontraditional Participants such as Youth, Women, and Minorities [7]

The problems associated with the transition of youths, women, and minorities into nontraditional occupations have been examined in some detail by manpower specialists. Unfortunately, the energy industries and their inherent occupational structures have not created large numbers of jobs for these groups. Historical employment patterns, the influence of organized labor, the location and character-

istics of the work environment, the nature of skills or training requirements, and a variety of other factors have posed entry problems
for youths, women, and minorities.

Nevertheless, energy occupations, particularly new occupations,
have few inherent characteristics that preclude the entry of these
groups. Therefore, a potentially fruitful area of research would focus
on the range of occupations that are accessible to women and minorities. The transition problems of youths are particularly difficult in
that the capital intensiveness and specialized-skills training present
major barriers. In addition, state laws frequently bar youths from
hazardous occupations, such as underground mining, heavy-equipment
operating, and working in electricity-generating plants.

Development of Labor Market Information Flows Utilizing CETA Prime Sponsors, State and Local Offices of the U.S. Employment Service, and Apprenticeship Programs [6]

One of the major findings of a labor market information study
conducted by the Human Resource Institute at the University of Utah,
for the Employment and Training Administration, region 8, was that
valuable sources of labor market information exist at the state and
local levels, but most of this information is not fully utilized. No
formal mechanism exists to channel quantitative and, more important,
qualitative information from local U.S. Employment Service offices
to state and regional decision makers.

The types of information that would be particularly useful in the
estimation of demand requirements include: where and when a firm
planned to initiate a new energy project; what type of expansion of existing plants could be expected, and how fast it would proceed; what
specific skills deficiencies faced specific firms; and what types of recruitment procedures firms used and their relative effectiveness.

The Comprehensive Employment and Training Act (CETA) prime
sponsors provide an all-inclusive coverage of the U.S. labor market.
Local Employment Service offices provide a less comprehensive,
though probably more intensive, coverage. Together they provide a
vehicle for data and information flows that may not be duplicated elsewhere in the federal-state system. Some effort should be made to
evaluate more intensively what sources are available and how an effective information flow can be established.

Approaches to Training [7-8]

The Office of Education in the Department of Health, Education
and Welfare and the Employment and Training Administration of the
Department of Labor have been involved in training and education for

several decades. Efforts under the Manpower Development and Training Administration, the Job Corps, CETA, vocational education, and a multitude of similar programs have provided the basis for intensive and extensive evaluations of what types of training and education are most job relevant. Nevertheless, there is definite evidence that the major issues have not been resolved.

Current debate focuses on higher education versus vocational training, on-the-job training versus institutional programs, upgrading versus retraining, and preapprenticeship in relation to apprenticeship programs. The energy industries and occupations are not unique in and of themselves (though some new occupation areas may emerge), but recent evidence suggests that not all types of preparation are equally effective from the employer's standpoint. For example, a recently completed labor market study in federal region 8 concluded that workers completing an instructional education-orientation program were not preferred to workers who had no training. An intensive study of different approaches to occupational training is required.

In our view, successful training is a key to effective labor market intervention strategies. Each training approach—on-the-job, institutional, vestibule, or apprenticeship—can play a significant role in energy occupations. However, present knowledge does not state what training mix is most effective.

Productivity in the Energy Sector [6-7]

Productivity has been rising in virtually every energy industry, with the exception of coal. In 1976, output per man-day averaged only 8.5 tons for underground miners. In 1972, it was almost 12 tons per man-day. Surface miners produced 25.50 tons per man-day in 1976, down from 36.33 tons per man-day in 1972.

There is wide variability in productivity within the two coal mining sectors. In Utah, for example, underground productivity ranged from a low of about 10 tons per man-day to over 50 tons per man-day. Size of operation, technology, type of coal seam, and other factors undoubtedly account for these differences. Still, because coal has such a vital role in the National Energy Plan, research into methods of improving productivity is warranted.

Identification of the Size and
Timing of Energy Projects [8]

Private energy firms operate in an environment of uncertainty characterized by changes in economic variables, legislation, environmental requirements, political forces, technology, and social factors. In addition, most firms are reluctant to release information that they feel is proprietary. As a consequence, obtaining information on the type, size, and timing of energy projects has been difficult. The Environmental Protection Agency is attempting to develop a tracking system built on licensing and permit procedures. The Bureau of Mines is developing a system to enumerate completely all prospective energy projects using a direct-employer contract. Furthermore, trade or professional associations in some industries are attempting to develop a listing of proposed energy development. Nevertheless, the fact remains that no comprehensive source exists that describes new energy initiatives or expansion plans.

The critical nature of this gap in the data base in obvious. If a reasonable approach to occupational training or mobility has any hope of succeeding, very specific information regarding the size, type, location, startup date, and length of development of energy projects must be obtained.

Development of a Long-Range Estimating Model for Energy Projects [8]

The national response to the energy problem will ebb and flow as the severity of the problem increases and decreases. Nevertheless, long-run trends will be established fairly early in the process as certain options are foreclosed. These trends may not be immediately obvious to all policy makers, and, more important, the implications of the trends may not be clearly discerned.

Many highly skilled occupations have lead times of eight or ten years. Rational planning does not permit the luxury of focusing on problems only when they are imminent. Bottlenecks in production, foreclosure of viable options, inefficient resource allocation, and a variety of other problems accompany efforts to plan only for the short or intermediate run.

The energy problem provides a clear example of how the lack of a contingency-planning capability can aggrevate a crisis. Therefore, we suggest that a separate, but administratively integrated, long-range energy/employment model be developed.

The Employment Impact of Energy Availability [10]

If the oil embargo taught the nation anything, it was that the United States is very dependent on foreign oil, and that this foreign resource is a critical part of our economic activity. The nation can struggle along with fewer Swiss watches or English sports cars and probably not experience dramatic changes in employment, consumption patterns, or gross national product, or an adverse international balance of payments. However, when the flow of foreign oil was reduced by 17 to 20 percent in late 1973 and early 1974 the economic imwas direct and immediate. Reverberations from the fourfold price change are still being felt.

The domestic impacts of the embargo have not yet been fully assessed, but three things are certain: GNP is lower today than it would have been without the embargo; unemployment is higher; and prices are higher. While several attempts have been made to quantify these changes, it is very clear that relatively little is known with any degree of certainty.[1] In fact, Robert J. Brown, regional administrator, Employment and Training Administration, U.S. Department of Labor, region 8, evaluated some of the recent work on the subject and concluded that "we are clearly still in the Dark Ages in terms of our capacity to analyze disruptive energy actions."[2]

What is particularly disturbing is that very little concentrated attention is being given to the problem. If another major disruption occurred tomorrow, this nation would be in no better position to respond effectively than it was earlier. In fact, with today's higher rates of unemployment and inflation, the United States is in a worse position. In addition, with our increasing dependence on foreign imports of oil, the United States is becoming progressively more vulnerable.

There is much that can and must be done in terms of assessing the 1973 embargo and, more important, setting up the machinery to deal with another interruption. All of the other energy issues and concerns would fade in importance if another embargo were to catch the United States unprepared. This issue deserves the highest priority.

The Employment Effects of the Proposed Energy-Storage Program [6]

Domestic reserves of all liquid and gaseous fossil fuels are rapidly dwindling. On the international level, new discoveries are approximately equal to withdrawals for consumption. (New reserves in the North Sea, Mexico, the Middle East, and elsewhere have added significantly to the world reserve base.) In any case, the United States must develop readily accessible petroleum deposits as contingent protection against supply interruptions. In addition, the International Energy Agency agreement, to which the United States is a signatory, requires development of a storage program. The nature of the storage program, while not precisely determined, encompasses the deposit of crude oil in salt domes or other relatively impermeable underground storage reservoirs. Oil can be extracted more quickly from these sources, in the event of a supply disruption, than it can from oil wells.

The strategic storage program may ultimately provide for the deposit of up to 2 billion gallons of oil in underground storage areas. The employment dimension of this process involves construction of storage facilities, transportation of petroleum reserves, and maintenance of storage vaults. The time frame of development will determine, to a large extent, what the employment impact will be.

While the strategic storage program does not entail development of new technologies or the expansion of domestic production, it is still a fairly important element in formulating a contingent-response strategy against oil shortages.

The Employment Impact of Energy Tariffs [5]

Most changes in tariff policy require a statement about the likely employment impact. Under the Trade Expansion Act, workers displaced by foreign competition can qualify for retraining and other types of aid. Assistance to displaced workers should continue, but other dimensions of the tariff policy, and its effect on employment, should be investigated. For example, the low cost of foreign petroleum, in terms of production and transportation, makes it possible for foreign governments to undercut any foreseeable market price of oil and gas flowing from the oil shale or coal-conversion industries. Hence, firms are reluctant to commit billions of dollars to the development of these resources.

The direct and indirect employment potential of the oil shale and coal-conversion industries is very large. Thus it seems reasonable to suggest that alternative tariff policies could have a substantial employment effect. However, the impact would most likely be geographically and industrially specific, and therefore might not have widespread importance.

The Geographic and Demographic Redistribution of Employment and Income Resulting from Energy Production and Use [6]

Americans worked shorter weeks after the embargo; they traveled less on their vacations; they bought smaller cars; they drove at slower speeds. Each of these changes occurred primarily because of the cost and availability of energy resources. Americans now spend relatively more on energy resources (home heating, air conditioning, gasoline, aviation fuel, coal) and therefore relatively less on other goods and services. The precise nature of this demand shift and its subsequent employment impact are important issues that are still poorly understood.

Certainly, energy shortages impinge on different areas or groups in a different manner. Examples of these regional differences include the concern in the North over natural gas for heat during winter months and the concern of farmers in the Midwest over the availability of fertilizer at a reasonable cost. The substitution of coal for fuel oil in electricity generation produces environmental, storage, and transportation problems.

The point, of course, is that energy production and use patterns are altered by supply restrictions, changes in technology, and changes in relative prices. The complex interaction of these factors as they affect employment and income distribution is an important set of relationships that deserves further investigation.

The Effects of Alternative Legislation or Policy Changes on Employment [7]

A variety of policy and legislative changes will develop in relation to energy production and use during the next decade. Each option has different employment implications. To assist in structuring policy initiatives, a consistent method of impact assessment must be developed.

As recent Office of Management and Budget (OMB) and Congressional Budget Office news releases suggest, there is wide disagreement on the employment dimension of alternative options. Reliance on the established national econometric models such as DRI, Chase Econometrics, and Wharton, without a clear basis for evaluating their respective strengths and weaknesses, places policy makers in a very precarious position. An employment standard that could be applied to alternative energy policies would aid lawmakers.

Integration of Research and Development to Avoid Overlap and Duplication [7]

One of the most disturbing aspects of the U.S. response to the energy problem has been our piecemeal approach to problem solving. Virtually every federal agency has a series of committees focusing on various aspects of the energy problem. Very likely, there is a vast duplication of effort and little sharing of information. As an example, the OMB has on file six or more separate economic-impact studies that were all made simultaneously within a single geographic area. All of the studies had similar objectives, research methodologies, and funding levels. This duplication or overlap problem appears to be the rule, not the exception. Furthermore, there is evidence that the vast majority of study results never impact the decision-making process because they are highly technical, voluminous, and closely guarded. Because so much money is spent on research and development, results must flow into a wider and more accessible information channel than at present.

Several federal agencies are developing in-house mechanisms to address their own information needs. However, the pervasiveness of the energy problem mandates the development of an information-flow model that will provide comprehensive, systematic access to research results.

Development of a Regional and National Labor-Supply Model [9]

The energy industries will probably never encompass more than 3 or 4 percent of the labor force. However, the workers required by this complex of subindustries range from highly trained nuclear physicists, engineers, and scientists to a variety of semiskilled workers in construction and maintenance.

There are current shortages of workers in many of the skilled trades (welders, electricians, tool- and diemakers, pipefitters) and in some skilled-technician occupations. With a stepped-up energy program, there will likely be widespread labor shortages throughout the energy sector.

Energy industries need a labor-supply estimation procedure that can be disaggregated to the state and substate levels with high levels of industrial and occupational detail. In terms of an employment strategy, this effort deserves immediate attention. In fact, to be of maximum usefulness, a labor-supply estimation technique should be developed simultaneously with a national labor-demand model.

Reviewing Foreign Energy/Employment Experiences to Determine Implications for U.S. Policy [6]

While the United States experienced its first real energy crisis in 1973, other countries have been meeting periodic crises for many decades; two of the more obvious examples are Great Britain and Japan. An intensive review of how these countries have responded to energy shortages, both in terms of expanding output and in terms of the impact on the overall employment picture, would be an important element in a comprehensive analysis of the U.S. experience. It may be possible to determine policy-response patterns, validate research findings, or suggest areas of further study. The institutional, economic, and political differences probably preclude development of precise parallels, but nevertheless, some attention should be focused on foreign experiences.

We may have omitted some research areas that will require attention, and there may be some disagreement on the priorities assigned to specific agenda items. Nevertheless, this agenda covers the core of energy/employment problems, and we hope that research projects will emerge from it.

The next chapter is a normal follow-through from many of the ideas outlined in the research agenda. Some items can be approached incrementally and on an individual project basis. Other subjects may require shifts in policy or new policy initiatives. A number of these policy suggestions are the subject of the next chapter.

NOTES

1. See, for example, Federal Energy Administration "Report to Congress on the Economic Impact of Energy Actions," as required by Public Law 93-274, sec. 51(d), FEA/B-76/351, Washington, D.C., May 1975.

2. Robert J. Brown, "Energy-Employment: The Critical Dependency in the U.S. Economy" (Paper delivered to the National Commission for Manpower Policy, San Francisco, October 19, 1976), p. 16.

9
POLICY RECOMMENDATIONS
RELATED TO ENERGY/
EMPLOYMENT RELATIONSHIPS

This study has attempted to bring together statements covering the current set of energy/employment problems, their causes, and methods that facilitate the understanding of the principal issues. Understanding energy/employment relationships is an essential part of dealing with energy problems. However, research alone is not sufficient. Because energy labor market problems influence nearly every aspect of the U.S. economy, it is important that the nation execute new policies to address these issues. Present approaches have not adequately met the problems created by rapid energy development or energy shortages. It is in the context of examining alternative policies that we advance the following macro and micro options.

MACRO POLICY SUGGESTIONS

Development of an Energy/Employment Model

The econometric models we have do not sufficiently assess the quality and quantity of the employment impact. The models have specification errors, use inadequate data, and aggregate at the state or the regional level. It is no wonder that human resources economists working at the firm or area level generally view the results of such models with some skepticism. Better models that are adequate for human resources policy formulation are needed. One such model has been developed by Boyd Fjeldsted at the University of Utah, and the preliminary results from this model appear promising. However, as every good human resources economist knows, one computer printout does not make an alternative future.

Consideration of the Complementarity of
Capital, Labor, and Energy

In formulating policy that affects basic, good producing industries, it is probably more important to think of labor, energy, and capital as complements rather than substitutes, even though some substitution may be possible. Jorgenson and Forrester contend that higher energy prices will force a significant substitution of labor for capital. However, this view seems to ignore not only the income effects of slower growth but also the real complementarity of energy consumption, life-styles, and labor force participation. In addition, some economists have calculated the employment effects of higher energy prices, but few have considered the loss in technological advances. There are long-term payoffs from the technological changes growing out of the substitution of capital for labor. The marginal costs and stream of benefits produced by these changes have seldom, if ever, been carefully analyzed. It seems appropriate to grant more thought to the technological implications of energy/employment tradeoffs.

Allocating R&D Expenditures to Assess the
Human Dimensions of Alternative Energy Policies

Given the present expenditure of $9-to-$10 billion for energy and energy-related research and development, the employment impacts of energy policies surely deserve thorough and continuous study. The importance of the human side of energy production and consumption (in relation to workers, employers, and consumers) requires that a predetermined percentage of energy R&D expenditures be earmarked for the analysis of these relationships. For instance, it is entirely possible that the coal industry, which is now in the early stages of a boom cycle, will end up in a bust within the next two or three decades. If this occurs, measures to mitigate the human effects must be taken. Changes in the mix of energy sources will cause labor market and consumption changes that should be foreseen.

Limiting the Uncertainty That Surrounds
Energy-Project Development

It makes little sense from an economic, social, or political standpoint to promote uncertainty leading to misallocations of resources, disruptions in the market, undue financial burdens on companies and communities, and, in many cases, the final abandonment of an energy project. Conflicting laws and regulations have plagued the energy sector and caused many energy projects in advanced planning stages to be sidetracked and abandoned. Utah's Kaiparowits project failed after 11 years and millions of dollars of planning; several oil shale projects in Colorado, Wyoming, and Utah have been initiated over the last three or four decades in a series of booms and busts; offshore oil and gas drilling and governmental coal-leasing programs have proceeded in a stop-and-start manner depending on public opinion.

In view of the costs that uncertainty entails, we recommend legislation that would prevent the controversy and indecision surrounding virtually every major project slated for development in the United States. We believe that energy projects should be subject to intensive study and should be developed in a manner that is environmentally safe, economically justifiable, and politically acceptable. However, there must be a statute of limitations imposed that will provide assurances that a reasoned decision will be forthcoming. Whether provided by the Environmental Protection Agency, the Bureau of Land Management, or any other federal or state regulatory agency, it is critical that a process be established to produce a "go" or "no-go" decision on energy development in a timely manner. While there may be no obvious way to reduce court challenges or shorten the decision process in the courts, it still seems to make eminent sense to rationalize and streamline the administrative process.

MICRO POLICY SUGGESTIONS

Improved Planning for Energy Boomtowns

Inadequate planning lowers the standard of living in a boomtown community and leads to a chaotic labor market. Community planners must anticipate an influx of construction workers, a rapid decline upon completion of the project, and a period of long-term stability with the advent of the operation and maintenance work force. CETA programs, particularly in relation to training, can help prepare the indigenous labor force for maximum participation in the energy-development process. It is important to develop a policy of controlled growth in communities impacted by rapid energy development.

A Formal Training Program
for the Coal Mining Industry

The Department of Labor, the unions that represent the coal miners, and the mining industry should establish a systematic, formal miner-training program.* As an alternative to a training pro-

---

*After this was written, the authors became aware of the Department of Labor-sponsored miner-apprenticeship program being developed in Utah.

gram, a two-track apprenticeship program (one track leading to opera-
tion-maintenance occupations, the other leading to foreman-supervi-
sory occupations) should be considered. The absence of a training process in the coal industry is a serious deficiency at a time when coal
demand is up and mine productivity is down.

Allocations to Low-Population/Rapid
Energy-Development Areas

The basic CETA Title I funding is based on previous allocations,
unemployment, and income criteria. The formula channels resources
into high-density areas and away from low-population ones. Consequently, there is little funding available for training programs essential to the energy sector, such as training in coal mining and in oil
and gas extraction. The Department of Labor should consider federal
funding for programs of this type that affect the national interest but
are necessarily located in nonurban areas.

Greater Emphasis on the Socioeconomic
Impacts of Power Plant Siting

Environmental issues have been overplayed, and socioeconomic
impacts have been neglected, in decisions regarding the siting of new
power plants. Large-scale energy projects could provide significant
employment in construction, operation, and maintenance, in areas of
high unemployment. Another consideration in the siting of power
plants is to choose areas where labor market adjustments may be ac-
commodated without the addition of large amounts of new social infrastructure. Of course, there are economic variables. In the short
run it may appear cheaper to build a coal-fired plant at the mine
mouth, but when the socioeconomic impacts are calculated, alternative siting may be less expensive. Policy on siting decisions is particularly important in view of the fact that power plant capacity may
double within the next 10 to 15 years.

Establishing an Interagency Task Force
to Improve Coal Mining Conditions

The coal mining industry has never been cited for exemplary
practices in training, health and safety, labor relations, turnover,
management, or employee morale. The coal miners' strike of 1977/78
should have taught the nation an invaluable lesson in how poor working
conditions in one vital industry can impact the entire economy. A
broadly based effort to improve working conditions, training, health
and safety, productivity, and labor relations in the coal mining industry should be a matter of high priority. However, the issues that affect working conditions and labor relations are complex. It may be

difficult to identify which government agency or organization within the industry could best improve the situation.

An interagency task force composed of representatives from organized labor, the coal industry, and government should be established to coordinate, assess, and facilitate efforts to address conditions in the coal mining industry. * At the present time there is no coordinated, comprehensive strategy to improve health and safety in the mines, structure better labor-management relations, or develop training or job-enrichment programs for miners. Coordination is important because the issues that affect working conditions are closely related. Training affects safety; safety measures impact productivity; and productivity is an important element in labor-management relations. Thus, problems in the coal mining industry cannot be dealt with in isolation.

The Development of a Data Task Force on Energy and Employment

We recommend the development of a data task force that will evaluate the current data-gathering system and suggest a specific strategy for improving the comprehensive data base and data flows.

As the analysis in this study proceeded, it became very obvious that two basic limitations exist: the first is that energy/employment data sources are poorly developed, and the second is that the available analytic tools, while not completely useless, leave much to be desired. A task force on data and analysis in energy/employment relationships should improve our abilities in these areas.

At various places throughout this study, we have suggested that appropriate modeling can provide insights into various phenomena. But even the best model will be worthless if the data that are put into it are inadequate. In and of themselves, models are rather sterile, uninteresting caricatures of the economy. They are unlikely to be very useful in decision making, or in influencing decision makers, unless good-quality data flows are available. The data provided at the local labor market level are meager at best. For instance, while it is easy to suggest that labor market analysts should attempt to estimate various employment multipliers, this task is an extremely difficult and uncertain one without an improved source of statistics. The point is that energy development intensifies our need to understand the labor market as thoroughly as possible.

---

*A presidential coal commission was initiated after this section was written.

SUMMARY AND CONCLUSIONS

As the subject matter of this study was pulled together and analyzed, it became increasingly evident to us that our motivations for
undertaking this effort were correct. We must be careful to point out,
however, that our criticisms are forwarded in the most positive
terms. It is fair to note that many individuals working in this most
difficult area have significantly advanced our understanding of energy/
employment relationships from what it was two or three years ago.
Individuals such as Bruce Hannon, Clark Bullard, James Benson,
Boyd Fjeldsted, Frank Hopkins, and Robert Brown deserve particular
credit. These individuals have taken either a policy or analytical interest in these issues and, by their perseverance, have pushed the
frontiers of knowledge outward. Much, however, remains to be done.

The United States and indeed the entire world must take the
energy problem more seriously. Part of the credibility problem has
been the exaggeration of the immediate energy crisis and the inability
of Congress and the administration to forge a comprehensive national
energy policy. In the next five or ten years, the energy problem will
be reflected in higher gasoline prices, astronomical electric bills,
but few, if any, actual energy shortages. Hopefully, as the national
energy policy emerges, a more orderly, equitable, and efficient
transition into the new energy era—an era of higher energy prices, a
shift to alternative energy sources, increased conservation, and, due
to these factors, improved efficiency—will occur.

While adequate evidence exists to suggest that energy and employment policies are complementary and in fact must evolve together,
we are concerned that few policy makers and researchers are focusing on the policy relationship. In a sense, an opportunity to significantly advance the state of the art of labor market analysis, to enhance the employment opportunities of millions of workers, to improve the distribution of income, and to utilize more efficiently all of
our resources may be slipping away. This is an important set of concerns that should not be permitted to be resolved by inaction or de-
fault.

There is no common thread through which energy/employment
relationships can be analyzed. These relationships are different in
some sense from most other economic or labor market phenomena.
Energy is probably more pervasive than any other factor input or
consumer item. Virtually nothing happens in contemporary society
without the expenditure of some form of energy. As a consequence,
macro analyses fall short of being an adequate tool for making public
policy. Even so, once we move toward a disaggregated economy, the
longstanding data and methodological problems that have plagued analysts for decades begin to arise. Our purpose was not to attack those

problems, but rather to attempt to identify the importance of energy/ employment relationships in the current energy policy debate. It would be presumptuous indeed to suggest that many answers exist. The interrelationships are so complex and pervasive that even the most powerful modeling and analytical techniques fall short. Our purpose was to point up the importance of these issues and suggest where research should be focused.

The tone of the study hopefully sparked optimism in terms of what can and must be done. We are aware that a number of modeling efforts have met with frustration, and failed, or have fallen far behind schedule. Others have been initiated with a grandiose scheme to answer all problems, but were subsequently scaled down. This should not be interpreted in any sense other than the recognition that the problem is extremely difficult and that only marginal improvements in our abilities to analyze and forecast are likely to occur.

As we view the evolution of energy and employment policies in the early 1980s, the dual attitudes of frustration and optimism prevail. Policy makers need and presumably want the best guidance available on energy/employment relationships. They are frustrated in most cases because of the absence of credible analysis, the conflicting results of analytical efforts, or the inability to adapt existing analysis to specific problems they face. Nevertheless, there is considerable room for optimism, in the sense that we now recognize that these problems exist, that they cannot be summarily resolved through analogy with previous studies, but that we know a great deal more today than we did two or three years ago. If this study has stimulated interest in this subject or provided insights into the nature of the problems involved, it has served its purpose.

APPENDIX
MEASURING LABOR REQUIREMENTS
FOR ENERGY INDUSTRIES

Defining the labor supply is one of the most difficult tasks that labor market analysts perform. An effective measure of labor supply must permit analysis by occupation, industry, location, and time. In addition, there must be some measurement of the available labor units, in terms of education and experience embodied by each worker. Presumably, as a worker increases his education, he acquires additional units of labor.

It is important to recognize that labor supply can be conceptualized as the available labor units. This may not be synonymous with the number of laborers. Employers and labor market analysts are naturally concerned with the relationship among capital, labor, the structure of production, and productivity. Except in situations where a necessary skill level is an absolute minimum, the employer will attempt to select the most potentially productive worker. This has several implications for the quantification of units of labor.

First, we may say that if the firm selects a person who possesses specified productive potential, other workers with similar attributes may be assumed to possess similar units of labor. Second, if we assume that the worker is paid the value of his marginal product, wages reflect labor productivity. All workers under a particular occupational wage rate may be assumed to possess the same units of labor. Third, some measure of educational attainment or years of experience can be used as an indicator of embodied labor units. The problems of commensurability are likely to be significant ones in this approach.

Before moving further into the elements of labor supply, it is essential to point out that the degree of participation by the various segments of the population will also affect the size and composition of the labor force. The labor force participation rate (LFPR) for a

FIGURE 7

Factors Determining Employment and Unemployment

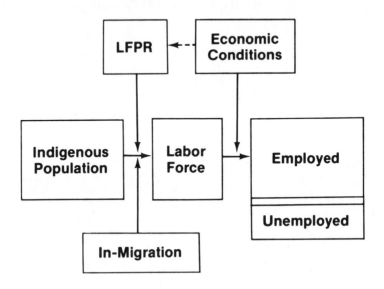

*Labor force participation rate.

Source: Compiled by the authors.

segment of the population shows what portion of that segment chooses to enter the labor force. Obviously, the LFPR will depend upon demographic characteristics, economic conditions, attitudes toward work, and a variety of other factors. A simplified schematic of these relationships is shown in Figure 7.

There are several reasons why measuring the labor market supply is so difficult. First, as noted earlier, the theoretical base of labor supply is poorly developed. In fact, there is no general theory of labor supply that has gained widespread acceptance by economists. We simply do not completely understand why people work, what motivates them, why they move or do not move, what role noneconomic factors play, and what training or skills-acquisition devices are most effective. In short, without a "road map" to guide empirical investigations, labor market analysts have been somewhat reluctant to work in this area.

Second, the components of the labor supply are not precisely defined and are thus difficult to measure. Every worker has a unique combination of skills, education, job experience, motivation, and economic needs that influences his or her performance on the job. A firm may decide that it needs a welder of a specific type. In response to that need, a large number of welders in the market may feel they have the required qualifications. However, as every employer will quickly point out, most of the individuals who consider themselves welders (whether due to education, experience, or whatever) are not qualified for the position. Yet, if a survey had been made prior to the demand change, the analyst would probably have found an adequate supply of welders available. The point is, of course, that in this case, a welder may not be a welder (at least, not in the firm's view). It is important therefore to be able to estimate with some precision what the supply components look like. The underlying problem, of course, is that we do not fully understand, or are unable to measure, the factors a firm uses to select or reject specific workers. In some cases, the most important criteria are noneconomic or unrelated to the job requirements.

Third, the supply of labor is sensitive to a variety of factors, including overall economic conditions, patriotism, wages and working conditions, individual economic circumstances, information flows, and family life cycles. It takes little imagination to recognize the difficulties of adequately incorporating these factors into the supply estimation.

Fourth, the supply side of the labor market is fragmented into a number of autonomous or quasi-autonomous sources. The most important sources of labor supply are educational institutions (high schools, universities, community colleges), apprenticeship programs, industrial training programs, in-migration, manpower-training programs, the armed forces, and self-training. To gauge total supply, each of these components must be estimated. They are not all of equal importance in a particular estimation process. (For example, if the estimation effort focuses on the supply of brain surgeons, presumably the analyst can ignore the self-training category.) However, for some occupational categories, all may play significant roles.

Fifth, geographic mobility also causes problems in measuring labor supply. Some workers are relatively mobile, while others require large net economic differentials to entice them to move. For example, older workers are generally much less mobile than younger workers. Married women are less responsive to economic differentials than are men. Expectations workers have in relation to the prospects of employment influence their willingness to move. If the prospects of finding a new job in an area are remote, a worker may feel compelled to move. If, on the other hand, conditions are improving, he may be inclined to wait it out.

A sixth problem involves the movement of workers into and out of the labor market in response to employment opportunities. We know, for example, that the supply of labor will increase during periods of economic expansion because secondary workers move into the market to absorb the excess of job openings. However, during tight markets workers may withdraw from the labor force because they believe no jobs are available.

In addition to these observations, there is the ever-present problem of adequate information. The dimensions of the information problem are implicit in some of the previously discussed problem areas. There are a number of qualitative aspects of the labor supply, such as attitudes, experience, and innate ability, that do not have easily identified quantitative proxies.

The somewhat protracted discussion of the nature of the labor-supply estimation process demonstrates that the problems are pervasive and very difficult. The estimation mechanism must confront all of the inherent obstacles. It is important to point out, however, that the problems do not preclude making meaningful estimates. On the contrary, most of the problems present difficulties at the estimation margin. In other words, only when attempts are made to make the estimates very precise does the analyst run into the full force of these problems. Reasonable estimates can be made that will provide useful policy guidelines. What is important is that the nature of the difficulties be clearly understood so that inappropriate policy recommendations can be avoided.

METHODS OF ESTIMATING ENERGY-MANPOWER REQUIREMENTS

The methodology used to estimate energy-manpower requirements is essentially no different from estimating manpower requirements for any identifiable segment of U.S. industry. If there is a practical element that differentiates energy-manpower estimating from other categories of estimates, it is the identification of occupational and industrial categories unique to energy industries. It may be argued, however, that there are no unique occupations in the energy sector. Consequently, to the extent that energy industries increase their demand for labor, they attempt to attract labor away from other industries. The activities involved in energy production are so pervasive that virtually every occupational category appears somewhere. For example, the scope of energy production includes exploration, facility construction, extraction, development or processing, maintenance, transportation, and ancillary activities. Each of these activities encompasses part of a larger industry and may not typically be identified as part of the energy sector. Further, the energy sector itself encompasses a variety of subindustries, such as gas, oil, coal, electricity, oil shale, geothermal, and solar, that display some unique industrial properties in addition to their general energy characteristics. A good example is the production of hydrocarbons for use in the chemical and plastics industries. It is reasonably clear that every industry is dependent upon the availability of energy as one of its principal inputs. Some rely more heavily on energy inputs than others, but virtually all industries are becoming more energy intensive. Reliance on secondary energy (electricity) is increasing, while usage of primary energy (direct consumption of coal, oil, or gas) is decreasing relatively. It is generally recognized that secondary energy sources are less efficient than are primary sources.

In any event, as energy production expands, additional labor inputs will be required. The techniques used to estimate the quantity and quality of these energy production changes are relatively rudimentary. At a minimum, the following information must be available to the labor market analyst. They are: a determination of when an energy project will be developed, and the sequence (timing) of development; the location of the project; the nature and size of the project, that is, a 5-million-ton-per-year surface coal mine or a 350-megawatt, coal-fired power plant; and an occupational-requirements table for the construction and the operation and maintenance phases of the project. The elements of the estimate are specific in terms of geography, time, industry, and occupation. The estimates can be quickly revised as more or better information becomes available.

The principal advantages of this direct-estimation process are that (1) the data are usually available to the analyst on a timely basis; (2) the techniques are relatively simple and can be replicated as conditions dictate; (3) the results are time and site specific; (4) the results can be easily interpreted; and (5) the estimates provide orders-of-magnitude requirements that should suggest the most likely labor market bottlenecks. The disadvantages are significant, however. They are as follows: (1) secondary and tertiary requirements are not forecast; (2) the data are project specific and not appropriate for aggregation over geographic areas or occupational categories; (3) it is difficult to extrapolate among types and sizes of energy projects, that is, employment-requirements tables usually pertain to specific size categories such as 1-, 3-, or 5-million-ton-per-year mines; and (4) there is no provision for changing technology.

The advantages and disadvantages of this technique are reasonably clear. The analyst can develop a set of estimates with little effort but must recognize their deficiencies. It is frequently useful to ask industry and union representatives to validate the estimates for a specific project before placing confidence in them. Even this procedure does not eliminate the first, second, and fourth disadvantages noted above.

The next level of analysis involves more than armchair estimates of employment needs. Virtually every firm will have developed estimates of the labor requirements for its particular plant. These estimates reflect the size and technological conditions more accurately. For example, the labor requirements for an underground coal mine using the conventional mining procedures will differ from a mine using long-wall techniques. Most requirements tables group all underground coal mines into one category.

The second technique involves a systematic sampling, or in some cases a complete enumeration, of the energy projects in a given area. Sampling procedures, data gathering and analysis, and construction of meaningful estimates are relatively standard artifacts of the contemporary analyst. Time, resources, and the degree of confidence required in the estimates will dictate the sampling or enumeration required. More important are the advantages and disadvantages of the technique. The advantages include the following: (1) this type of analysis more accurately reflects the actual conditions in the industry and in the participation firm, that is, size, technology, resource availability; (2) the results are time and site specific; (3) the estimates are relatively ambiguous; and (4) the estimates are likely to be more readily acceptable to the firms and unions, and in general the analyst will have more confidence in them. The disadvantages are that (1) the process is more expensive and time consuming; (2) not all firms prepare specific requirements tables or make them available; (3) replication may not be feasible if the procedure proves costly; (4) the technique still misses the secondary and tertiary effects; and (5) the problems of aggregation remain.

At this level of analysis, the critical problem then becomes: Can enough additional information be generated by this approach to justify the required time and resources? Most analysts would argue that this is a better technique in some respects, but a number of serious deficiencies remain. If the analyst needs fairly precise demand data for a specific firm or geographic area, this approach may be appropriate. If conditions are changing rapidly, or if only orders-of-magnitude estimates are needed, the first approach may suffice. There are no clear-cut guidelines that will assist the analyst in making his decision on the best approach. Resources, time, ability of the analyst, availability of data, use of the estimates, and a number of other considerations are important.

The third method of estimating energy-manpower requirements may take the form of one of several basic approaches. Space limitations do not permit a complete description of techniques, but adequate references are available for the interested reader. In brief, there are four approaches, some of which may be integrated with each other. The four methods are the shift-share technique,[1] input-output analysis,[2] econometric or regression techniques,[3] and economic-base studies.[4] While the techniques involved in these approaches vary widely, their advantages and disadvantages in relation to the two estimating methods discussed previously fall in the same generic categories.

In general, these approaches can be used to estimate employment needs at various points in time by changing the values of the variables and presumably obtaining a "simultaneous solution" that yields the total employment requirements at a point in time. Further, these approaches possess distinct advantages because (1) they include historical and contemporary forces that impinge on the labor-demand function; (2) they can easily be rerun using better or different data and assumptions; and (3) more factors can be incorporated into the equations to reflect actual market, economic, and industrial conditions.

The disadvantages, also impressive, are that (1) the analyst usually must possess rather sophisticated analytical capabilities; (2) the data may not be adequate to justify the more powerful analytical techniques; (3) users may not understand the techniques or the results and therefore may place little confidence in them; and (4) they are likely to be relatively expensive systems to develop and maintain.

In any case, the more powerful techniques are being developed, and, in fact, there are a variety of efforts in progress that incorporate two or more of the basic approaches into the analytical framework. For example, Fjeldsted at the University of Utah is using a hybrid input-output/economic-base model to estimate regional energy manpower requirements. In addition, a number of the large, macro models, that is, DRI, Wharton, and Chase Econometrics, which are essentially econometric models, include an input-output submodel for various estimating purposes.

The proliferation of modeling efforts has resulted in a virtual landslide of conflicting estimates of the employment impact of the National Energy Plan and its various components. Everyone recognizes that while the aggregate estimates are interesting, their most important function is to serve as control totals for modeling at the regional, state, and local levels. It should also be recognized that modeling at the subnational level is extremely difficult, and that estimating errors may be large. The United States needs a generic model of subnational economics that incorporates local coefficients and produces localized estimates.

However, there are numerous problems related to a modeling effort of this type. In most instances, there is an absence of localized data that are timely and accurate. Also, local conditions can change rapidly as large numbers of outsiders move into an area. Another problem is that due to the tactics of environmental groups, pressures from local political leaders, and the vagaries of the legal process, the timing of a project's startup and development is frequently uncertain. Furthermore, models are expensive to develop and maintain, and local planners or political leaders frequently do not have the expertise or perception to use the output from these efforts.

NOTES

1. For a general description of this method, see U.S., Department of Commerce, Bureau of Economic Analysis, Area Economic Projections 1990 (Washington, D.C.: Government Printing Office, 1974), pp. 1-10 and A1-A5. For a more complete description, see David Houston, "The Shift and Share of Regional Growth, A Critique," Southern Economic Journal 33, no. 4 (April 1967): 577-81.

2. See William H. Miernyk, The Elements of Input-Output Analysis (New York: Random House, 1965); and Boyd Fjeldsted and Iver Bradley, "The Utah Input-Output Project," Utah Economic and Business Review 35, no. 10 (October 1975).

3. See Elchanan Cohn et al., The Bituminous Coal Industry: A Forecast (University Park: Institute for Research on Human Resources, Pennsylvania State University, 1975).

4. Charles M. Tiebout, The Community Economic Base Study, Supplementary Paper no. 16, Committee for Economic Development, Washington, D.C., December 1962.

ANNOTATED BIBLIOGRAPHY

Association for University Business and Economic Research. "Socio-Economic Impact of Electrical Energy Construction." Proceedings of the First AUBER Energy Workshop, Snowbird, Utah, August 22-26, 1977.

The association assembled a group of national experts on social and economic impact, including Dr. John S. Gilmore of the Denver Research Institute, Dr. F. Larry Leistritz from North Dakota State University, and Dr. Arthur F. Mehr from the Los Alamos Scientific Laboratory, to discuss this subject in relation to the construction of electric power plants. The papers covered a wide spectrum of issues, from model building to impact assessment to impact assistance.

Baker, Joe Garrett. Labor Allocation and Western Energy Development. Monograph no. 5. Salt Lake City: University of Utah, Human Resource Institute, 1977.

This study addressed the mechanisms of labor allocation in small, rural energy-impacted communities in the Rocky Mountain region. Central to the discussion was the labor-allocation process, labor market constraints, and policy or market-intervention strategies. Three energy-impacted communities were intensively studied. The principal impacts occurred in the secondary labor markets.

Bechtel Corporation. "Direct Requirements of Capital, Manpower, Materials and Equipment for Selected Energy Futures." Report prepared for the Energy Research and Development Administration, under contract no. E(49-1)-3794, San Francisco, April 1976.

Bechtel Corp. used its Energy Supply Planning Model to convert alternative energy futures into resource-requirements schedules. Alternative assumptions were used to test the sensitivity of various resource requirements.

Berndt, Ernst R., and David O. Wood. "Technology, Prices and the Derived Demand for Energy." Review of Economics and Statistics 57, no. 3 (August 1975).

Berndt and Wood examined the substitutability among energy and nonenergy resources. They found that "energy and labor are slightly substitutable," whereas "energy and capital are complementary." They argued that lifting ceiling prices on energy resources would tend to reduce the capital and energy intensiveness of producing a given volume of output and would increase the labor intensiveness.

Bezdek, Roger H. Long-Range Forecasting of Manpower Requirements: Theory and Applications. New York: Manpower Planning Committee, Institute of Electrical and Electronics Engineers, Inc., 1974.

This study represents a reasonably complete description of the state of the art of forecasting, utilizing an input-output model for estimating manpower requirements. The description is accessible to individuals with some background in quantitative economics.

Brown, Robert J. "Energy-Employment: The Critical Dependency in the U.S. Economy." Paper presented at the National Commission for Manpower Policy, San Francisco, October 14, 1976.

This paper analyzes the employment perspective of energy production with specific emphasis on the Rocky Mountain region. The author argues that the energy/employment interface has not received enough attention, and that the United States may experience labor shortages in the energy sector without careful preparation.

Bullard, Clark W. "Energy and Jobs." Paper presented at the University of Michigan Conference on "Energy Conservation—Path to Progress or Poverty?" Ann Arbor, November 1-2, 1977.

The author argues that energy conservation and increasing employment are not mutually exclusive goals. In fact, he makes the case that they are complementary goals. Further, "convincing arguments can be made that energy conservation can create jobs that are much more satisfying and fulfilling than the ones it eliminates." He argues that the rapid changes in technology probably preclude long-range employment estimating. He suggests that "short-term analysis may be the only one that can yield believable results."

Burggraf, Shirley P. "Energy: The New Economic Development Wildcard." Paper prepared for the White House Conference on National Balanced Growth and Economic Development, under a grant from the Economic Development Administration, Washington, D.C., January 1978.

This paper examines the potential role of energy in regional economic development. Generally, "it seems that energy is likely to have a converging effect on subnational development in that it is tending to push more states toward the average per capita income than away from it. " Burggraf concludes that "energy policy and economic development policy cannot be formulated independently but must instead be coordinated if the goals of both are to be served. "

California Public Policy Center. Jobs from the Sun: Employment Development in the California Solar Energy Industry. Los Angeles: CPPC, February 1978.

This study develops a solar scenario for California that will produce over 375,000 jobs per year, $41.2 billion in increased personal income, and $51.1 billion more in gross state product between 1981 and 1990.

Cogan, John M. , Bruce Johnson, and Michael P. Ward. "Energy and Jobs: A Long Run Analysis. " Original Paper 3. Los Angeles: Institute for Economic Research, UCLA, July 1976.

Cogan et al. examine the "long-run implications of alternative energy policies. " Their study is a theoretical analysis of alternative policies as opposed to a forecasting effort. Two primary conclusions emerge from their analysis: efforts to raise the price of energy domestically will result in increases in employment—that is, labor and energy are substitutes; with the exception of a moderate tax on crude oil or complete decontrol of the domestic oil industry, policies that reduce the dependency of the United States on imports reduce real per capita GNP.

Cohn, Elchanan, et al. The Bituminous Coal Industry: A Forecast. University Park: Institute for Research on Human Resources, Pennsylvania State University, 1975.

The book's title is somewhat misleading in the sense that the subject matter is essentially restricted to employment in the bituminous coal industry. The authors use an econometric model to estimate labor demand and a demographic-mobility model to estimate labor supply in this industry. The conclusion generated by the model is that there will be no labor shortages in the bituminous coal industry. "If anything, the results of this analysis indicate the more likely possibility of surplus, especially in the West. "

CONSAD Research Corporation. "Review of Employment/Energy Economic Analysis Methods. " Report prepared for the U.S.

Department of Labor and the U.S. Department of Energy, Washington, D.C., January 1979.

This report analyzes two Department of Labor models and six models developed under the auspices of the Department of Energy. The report "discusses the start-of-the-art in DOL-DOE models dealing with energy-employment impacts." The study is based on the assumption that considerable capability currently exists and that efforts should be made to expand that capability rather than initiate new modeling efforts.

Contractors Mutual Association. Projecting Construction Manpower Requirements: A Guide to Methods and Sources for Estimating Future Demand. Washington, D.C.: CMA, March 1975.

This CMA study is an effort to develop a methodology for estimating construction-manpower requirements, by occupation, at the state and substate levels. The objective of the study was to provide local planners with procedures for making long-range projections of labor demand that will provide enough lead time to recruit, train, or retrain a suitable work force.

Council on Environmental Quality, Executive Office of the President. The Good News about Energy. Washington, D.C.: U.S. Government Printing Office, 1979.

The council believes that lower growth in energy use, without adverse impacts on the quality of life, is possible and in fact is occurring. "The technology to increase greatly the productivity of the U.S. energy system is at hand." If productivity increases, as the council feels it can and must, "total U.S. energy use need not increase greatly between now and the end of the century, perhaps by no more than 10-15 percent."

Donnelly, William A., et al. "Estimating a Comprehensive County-Level Forecasting Model of the United States—READ." Washington, D.C.: Federal Energy Administration, rev., August 1977. Mimeographed.

The Regional Energy, Activity, and Demographic (READ) Model is an effort to assess the impact of energy policies on regional, state, and local economies. The READ Model is an econometric model that will ultimately contain about 2,500 equations. It was sequentially developed and was completed in the spring of 1978.

Energy Sector Office. Federal Interagency Construction Task Force: Energy Sector. Washington, D.C.: U.S. Department of Labor, January 1977.

This report contains a basic description of the Construction Manpower Demand System (CMDS). A variety of important appendixes are included that describe the methodology used in data collection, utilization of Contractors Mutual Association reports, and the parameters of the Civilian Enriching Requirements Program.

Environmentalists for Full Employment. Jobs and Energy. Washington, D.C., Spring 1977.

The Environmentalists for Full Employment is a group that argues that the protection of our environment and full employment are not mutually exclusive goals. In fact, they are complementary goals. This study brings together a series of studies that support the economic efficacy of conservation and small-scale energy development.

Executive Office of the President. The National Energy Plan. Washington, D.C.: U.S. Government Printing Office, April 29, 1977.

This document outlines the origins of the energy problem; the principles, goals, and basic strategies of the National Energy Plan. Each of the principal energy elements is outlined in detail. The role and interaction of various governmental levels are defined. The study concludes with a statement about the plan's impact on U.S. citizens and the prospects for the future.

Federal Energy Administration. Labor Report. Project Independence Blueprint Final Task Force Report. Washington, D.C.: U.S. Government Printing Office, November 1974.

This report was prepared by the Department of Labor for inclusion in the "Independence Blueprint." There are some useful baseline data for interregional analyses. It contains a variety of occupational profiles for some of the major energy industries. Due to the rapid change in the energy sector, much of the data are outdated.

_____. Report to Congress on the Economic Impact on Energy Actions, as required by Public Law 93-275, Section 18(D), FEA/B-76/351. Washington, D.C.: U.S. Government Printing Office, May 1976.

This report focuses on the impact of the 1973 embargo, and the resultant energy price increases, on the U.S. economy. The methodology was to simulate what actually happened in 1975 and compare these results with what would have happened without the embargo. During the last quarter of 1975, GNP was $66.9 billion lower, consumer expenditures were $37.7 billion lower, and investment expenditures were down about $30.2 billion. Inflation was 2.2 percentage points higher; unemployment was 2.5 percentage points higher.

_____. Short-Term Microeconomic Impact of the Oil Embargo: October 1973-March 1974. Office of Economic Impact Analysis. Washington, D.C.: U.S. Government Printing Office, n.d.

This report focused on energy-use patterns in U.S. industry and how they shifted as a result of the 1973 embargo. Attempts were made to determine how industry adjusted to energy shortages. These adjustments were reflected in output, prices, and changes in final demand.

Hannon, Bruce M. Energy, Growth, and Altruism. Urbana: Center for Advanced Computation, Energy Research Group, University of Illinois, October 1975.

This study won the 1975 Mitchell Award for Hannon. He attempted to identify the human reactions to energy shortages. Three dilemmas were described to show that there will be considerable resistance to reduced energy use. The dynamics of real wage reductions, of decreases in total income, and of redistributing income to conserve energy militate against significant reductions in energy usage. Energy rationing is the complete solution; a tax on energy, as it leaves the resource base, is a partial solution.

_____. "Energy, Labor, and the Conserver Society." Technology Review, March/April 1977.

This article was one of the first to suggest that labor and energy may be substitutable in the production process. Hannon shows how society can simultaneously decrease energy use and increase employment. He provides a ranking of changes that will result in the largest employment increase per energy unit saved.

Kramer Associates, Inc. Determination of Labor Management Requirements to Meet the Goals of Project Independence. Final report submitted to the Federal Energy Administration. Washington, D.C., August 27, 1975.

This study focused on the employment changes in the coal industry. The key conclusions were: (1) coal mine productivity should reverse its downward trend in 1976/77; (2) between 1974 and 1985, coal mine employment should increase by 52 percent; (3) there will be changes in the occupational composition of the industry; (4) there are likely to be training problems in some occupations; and (5) there are potential problems in obtaining experienced workers in western surface mines.

_____. Manpower Requirements and Availability for Projected Coal Production in Ten Eastern States to 1985. Report submitted to the Federal Energy Administration. Washington, D.C., September 8, 1977.

West Virginia is expected to have a deficit of 86,230 miners by 1985. Eastern Kentucky and Pennsylvania are also expected to experience small shortages, 2,591 and 7,783, respectively. The remaining coal-producing states in the East are not expected to have problems obtaining adequate manpower. The report concludes that "unless steps are taken to encourage miners from other areas such as Tennessee, Virginia, and Ohio, West Virginia will have a substantial problem by 1980, continuing to 1985, in securing the manpower required to meet the projected production goals."

Mead, Walter J. "An Economic Appraisal of President Carter's Energy Program." Reprint Paper 7. Los Angeles: International Institute for Economic Research, UCLA, September 1977.

Mead briefly summarizes a half century of federal energy policies, and shows that for the most part they were a "conflicting set of expedient measures." There are both commendable and questionable policy proposals in the president's program. Mead suggests that the government intervention in the energy business had disruptive effects and therefore should be restricted to situations with clearly demonstrated significant externalities. The market should be the mechanism for allocating scarce resources among competing ends.

Medvin, Norman. The Energy Cartel: Who Runs the American Oil Industry. New York: Random House, 1974.

The focus of this study is the interrelatedness of government and business in energy production and distribution. Medvin argues that energy companies almost dictate policy or regulations that impact their operations. A sequel was published in 1975 entitled The Energy Cartel: Big Oil vs. the Public Interest.

Mountain West Research, Inc. Construction Worker Profile. Billings, Mont.: Old West Regional Commission, December 1975.

This study attempted to collect and analyze data on large construction projects and their related work forces to determine how communities reacted to these projects and the types of impact that occurred. Fourteen projects in seven Rocky Mountain states were included in the project; over 3,000 construction workers responded to the questionnaire. Labor mobility, sociodemographic characteristics, community attitudes, expectations, and other factors were analyzed.

National Petroleum Council. "Availability of Materials, Manpower, and Equipment for the Exploration, Drilling, and Production of Oil—1974-1976." Report of the National Petroleum Council's Committee on Emergency Preparedness, Washington, D.C., September 1974.

The National Petroleum Council examined the resource requirements for the oil industry between 1974 and 1976. Some longer-run trends were provided to 1985. A number of material constraints were identified, but manpower was not believed to be a critical constraint overall.

Neilsen, George F. "Coal Mine Development and Expansion Survey . . . 617.3 Billion Tons of New Capacity 1977 through 1985." Coal Age, February 1977.

A survey by McGraw-Hill of the coal industry's expected coal mine expansion showed that about 617 million tons per year of new capacity will be added by 1985. About two-thirds of the increase will occur in surface mining. Most of the increased surface capacity will occur in the West; most of the underground increases will be in the East.

Office of Naval Research. "The Economic Impact of an Interruption in United States Petroleum Imports: 1975-2000." Paper prepared by the Center for Naval Analyses, AD-A010-914, November 1974.

This study focused on the potential economic impact of the interruption of petroleum imports during the period between 1975 and the year 2000. An input-output model of the United States was used to estimate the total impact. The paper concludes that U.S. vulnerability to petroleum disruptions will increase over the next 25 years and that our dependency on Middle East imports will concurrently be rising.

Oil, Chemical, and Atomic Workers International Union. OCAW Energy Policy. Denver: OCAW International Executive Board, May 1977.

The OCAW energy policy is based on the assumption that "energy is not a problem of geology. Nor is it a problem of economics. Energy is a political problem." The OCAW's perspective on energy policy is particularly important because this union has over 70,000 members in the oil, gas, and nuclear industries and is therefore very sensitive to the structure and capability of the energy sector.

PRC Data Services Company. "CMDS Conceptual Design." Report prepared for the U.S. Department of Labor, McLean, Va., December 10, 1976.

This report was one of the first complete descriptions of the then Construction Manpower Demand System (CMDS). It is an extensive description of methodology, data sources, and work assignments related to the CMDS.

Rall, Jane E. Energy-Related Scientists and Engineers: Statistical Profile of New Entrants into the Work Force, 1976. Oak Ridge, Tenn.: Oak Ridge Associated Universities, ORAU-147, October 1978.

This is the most recent in a series of studies of scientists and engineers in energy-related occupations. While the analysis is thorough and well written, there are no apparent surprises. Women and nonwhites are still underrepresented in the energy-related group. Those entering the energy sector are usually employed by private industry, and they generally command a higher median salary than do all recent graduates in the occupations.

Susskind, Lawrence, and Michael O'Hare. "Managing the Social and Economic Impacts of Energy Development." Summary report: Phase I of the MIT Energy Impacts Project. Cambridge: Massachusetts Institute of Technology, Laboratory of Architecture and Planning, December 1977.

This summary report outlines in a complete and clear manner what causes boomtowns, who is responsible for preventing boomtowns, or assisting communities under stress; and summarizes several impact-assistance programs that have occurred in the Rocky Mountain region. This is a well-written report that will provide those who are unfamiliar with the subject a short course on social and economic impacts.

Tiebout, Charles M. "The Community Economic Base Study." Supplementary Paper no. 16. Committee for Economic Development, Washington, D.C., December 1962.

Tiebout develops the methodology for conducting an economic-base study. The description is set forth in relatively elementary terms and can be readily understood by the nontechnician.

U.S., Department of Labor. "CMDS Workshops." Knoxville, Tenn.: DOL, November 4, 1977. Mimeographed.

This document contains a series of computer runs on regional estimates of energy-sector employment. Of most interest is the technical description of the Phase II econometric forecasting procedures developed in the Construction Manpower Demand System.

_____. "Forecast of Manpower Requirements for Electric Power Plants." Preliminary report. Washington, D.C., October 1977.

Some of the first computer runs related to employment in the energy-facility construction sector are presented in this report. The data are preliminary and not for official publication. They do show, however, the regional profiles of employment by major craft group.

U.S., Department of Labor, Employment and Training Administration, region 8. "Regional Industrial-Occupation Labor Demand Model Description." Denver, n.d. Mimeographed.

The Bureau of Economic and Business Research at the University of Utah has worked with the DOL's Employment and Training Administration to construct an innovative energy/employment modeling program. This report outlines the methodology, data sources, conceptual issues, and related information. It also contains data from an early run.

U.S., National Committee of the World Energy Conference. A Sensible Energy Policy Now: Today's Challenge to Meet 21st Century Needs. National Energy Forum V. Washington, D.C.: U.S. National Committee, May 23-24, 1977.

A variety of distinguished experts and public officials presented their views on the NEP and its relationship to the energy problem. Included in the list of contributors are: Governor Edwin Edwards, Louisiana; Dr. Alfred E. Kahn; and Senator Henry M. Jackson. The articles provide a well-rounded perspective on the energy problem from both the public and private sectors.

INDEX

absenteeism, 22, 27, 31, 32, 87, 88
advanced planning, 30
Alabama, 38
alcoholism, 23
alternative: energy sources, 3, 11-12, 15, 53-54; technologies, 6-7, 51, 63-64
Appalachia, 86
apprenticeship, 11, 90-91, 100-1
attendance, 28

balkanization, 59
Bechtel Corporation, 29, 31, 75
Benson, James, 68-69, 103
best available control technology, 38
bioconversion, 8, 34, 51, 67, 71; industry, 2
biomass conversion, 63, 65
blasters, 15
boilermakers, 50
boilermaker-welder, 45, 50
booms and busts, 99-100
boomtown, 21, 22-25, 27, 28-33, 86, 89, 100
breeder reactors, 8
Brown, Robert J., 93, 103
Bullard, Clark, 103
Bureau of Economic Analysis (BEA), 78-79
Bureau of Labor Statistics (BLS), 50, 64, 77, 78-79
Bureau of Land Management, 100
Bureau of Mines, 92

California, 64, 67-68; Geysers, 64; Public Policy Center (CPPC), 67

capital: -intensive projects, 58, 65; markets, 71; substitution of labor for, 99
carpenters, 45, 70
Carter, President, 4
Center for Advanced Computation, 56, 64, 69
Chase Econometrics, 75, 95
Clinch River Breeder Reactor Project, 58
coal, 5, 12-14, 28-29, 34-35, 50, 53, 63, 76, 91, 99, 100-2; bituminous, 15, 23; -burning states, 5; conversion, 15-16, 21, 64-65, 94; -fired electric generating plant, 25-26; gasification, 51, 78; liquefaction, 17-18, 51; low-sulfur, 38; mining, 8, 78, 89
Colorado, 80, 99-100
complementarity, 53, 99
Comprehensive Employment and Training Act, 90, 100-1
Congressional Budget Office, 95
conservation, 6, 52, 64; solar, 68-69
construction: of fossil-fueled plants, 42; of nuclear plants, 42; power plant, 46, 78
Construction Labor Demand System (CLDS), 42, 75-77
consumption, 3; per capita electricity, 46
Contractors Mutual Association, 76-77
control totals, 75
costs, psychic, 31
Council on Environmental Quality, 64
crime, 22-23, 24, 27, 30

ABOUT THE AUTHORS

WILLIS J. NORDLUND is the Special Assistant to the Under Secretary of Labor, U.S. Department of Labor.

Dr. Nordlund's interest in labor market policy and human resources development began in the mid-1960s when he entered the doctoral program at the University of Utah. He began the intensive examination of energy and employment relationships in 1974, while serving as the regional economist for the Employment and Training Administration, U.S. Department of Labor, in Denver.

Dr. Nordlund received his Ph.D. from the University of Utah in 1972. His dissertation research focused on an investigation of the economic impact of computer-assisted placement systems in the United States Employment Service. He has held academic positions at West Virginia University, Ball State University, Oregon State University, and Denver University.

Dr. Nordlund has published numerous articles in such professional journals as Industrial Relations, Arbitration Journal, Monthly Labor Review, and the Labor Law Journal.

R. THAYNE ROBSON is the Executive Director of the Bureau of Economic and Business Research, and Professor of Management and Research Professor of Economics, at the University of Utah.

He is a native of Utah and received his B.S. and M.S. degrees at Utah State University and did graduate work at Cornell University. In addition to the University of Utah, Professor Robson has taught at Harvard University and UCLA and has worked for the Brookings Institution.

Professor Robson has served as Executive Director of the President's Committee on Manpower in Washington, D.C., and as senior staff economist on the National Commission on Technology, Automation and Economic Progress. In addition to his teaching duties, Professor Robson has been a consultant to many state, local, and federal government agencies.

Among his current research interests are human resources planning for the energy industry, as well as a continued interest in monitoring the growth and outlook for the Utah economy and surrounding areas. He has contributed numerous publications to economics, human resources, and industrial relations literature.